Dams and Development in India

Dams and Development in India

Arun Kumar Nayak

RAWAT PUBLICATIONS
Jaipur • New Delhi • Bangalore • Guwahati • Kolkata

ISBN 978-81-316-0780-0

Published by
Prem Rawat for **Rawat Publications**
Satyam Apts, Sector 3, Jawahar Nagar, Jaipur 302 004 (India)
Phone: 0141 265 1748/7006 Fax: 0141 265 1748
E-mail: info@rawatbooks.com
Website: www.rawatbooks.com

New Delhi Office
4858/24, Ansari Road, Daryaganj, New Delhi 110 002
Phone: 011 2326 3290

Also at *Bangalore, Guwahati* and *Kolkata*

Typeset by Rawat Computers, Jaipur
Printed at Nice Printing Press, New Delhi

Dedicated to the
Displaced People

Contents

Preface

Dams are considered as an agency of development for its contribution to flood management, hydropower production, irrigation, navigation and supplying water for urban and industrial needs. While many have benefited from the services dams provide, their construction has led to many significant social and human impacts, particularly in terms of environmental costs and human displacement. Compulsory displacement that occurs for development reasons, embodies a perverse and intrinsic contradiction in the context of development and it raises the major issues of social justice and equity.

In this context, this book discusses on dams and its dynamics of development. Looking at the history of dam constructions in the world, although, a large number of dams were constructed in the past, especially, in the developed countries, the popularity of Multi-Purpose River Valley Development plan (MPRVD) took place in USA with the establishment of Tennessee Valley Authority (TVA) and this model was replicated by many nations of the world. Following the TVA model, dams were constructed on almost all the big rivers of the world. However, empirical studies have shown that many a projects have failed to fulfil their intended objectives. This raises the question of the rationality of TVA and its executions in other parts of the world.

The book further describes that the genesis of TVA in USA was connected to the particular socio-economic and political situations of 1930s. This period was marked by severe economic recessions and TVA was established in 1933 in order to immediate stimulation of employment in the region. The era was also marked by the rapid decline of the European colonial powers and the emergence of the United States of America as the new hub of capitalist dynamism. Many governments including India adopted MPRVD immediately during their post-colonial phase, with the help of American experts, for its construction to speed up their economic development. However, in reality, these projects remain far away from fulfiling such intended development. It can be said that, the MPRVD scheme was inextricably linked to the political economy of decolonization and nationalism rather than development. The integrated river basin management in India was largely promoted by many scientists, engineers and statesmen, inspired by the United States.

This book has vividly described the account of MPRVD on river Mahanadi (Hirakud Dam). The empirical scrutiny has shown that, none of the objectives of the dam were fulfiled even after fifty years of its completion. On the contrary, it has brought massive social misery by submerging 325 villages and displacing about 1,00,000 people. In spite of bringing about development in the region, it has now brought uneven development in the command area and several environmental consequences in the locality. Thus, conflicts are taking place among various stakeholders. The study shows that the model of TVA was blindly executed on the river Mahanadi without understanding the local feasibility. Even though the proposition was opposed by an engineer (M.G. Rangaya) in his report, the government did not pay any attention to his suggestions. Although protests and movements were made by local people, it was suppressed in the name of nation-building and nationalism in the early years of independence.

Acknowledgements

This academic endeavour would not have been possible without the help of a number of people, who, with their kind co-operation, active supervision and steadfast support have made this research work a reality. I am extremely grateful to my esteemed Professor Indra Nath Mukherji whose keen supervision, able guidance, intellectual ingenuity, scholastic co-operation and avid suggestion helped me in completing these academic exercises.

I am indebted to the staff of Jawaharlal Nehru University, New Delhi; South Asia Foundation Library, New Delhi; Centre for Science and Environment Library, New Delhi; Indian Social Institute Library, New Delhi and Indian Institute of Public Administration Library, New Delhi who have rendered active co-operation in the collection of books, articles, documents, records and relevant information.

I extend my gratitude to Mr Prafulla Samantara (President of Lokshakti Abhiyan, Orissa); Professor Durga Nayak, Sambalpur; Er Karunakar Supakar, Former Chief Executive Engineer, Orissa Hydropower Corporation Ltd. Burla; Mr Bhakta Prasad Nanda, President of Hirakhand Nagarik Parishad, Sambalpur; Mr Ashok Pradhan, President of Paschim Orissa Krushak Sangathan Sammanuya Samity; Mr Lingaraj, President of Orissa Krushak Sangathan Sammanuya Samity; Er Janki Mohapatra, Chief Executive Engineer, Department of Irrigation, Hirakud Dam Division, Burla; Mr Ranjan Panda, Director of Manav Adhikar Seva Samity, Sambalpur; Mr R.N. Nanda, Hirakud Land Organization Officer, Collector Office, Sambalpur; Dr Rajat Kujur,

Lecturer of Political Science in Gangadhar Meher College, Sambalpur; Dr Tripurai Narayan Pati, Lecturer of Political Science in Gangadhar Meher College, Sambalpur and my friend Mr Murali for their active co-operation during the period of my fields study for Hirakud Dam.

I am extremely thankful to all my friends Arup, Bharat, Santesh, Akshya, Hari, Pritish, Subodh, Anupam, Chandan, Subas, Anjan, Sarada, Laxmidhar, Gadadhar, Arun, Gyan, Manoj, Sangram, Maluaji, Beraji, Fanta Da and my good friend Late Shiv Narayan for their constant moral support and immense encouragement throughout the study. I am also grateful to my colleagues and Principal-in-charge, Dr Haradhan Saha for their support and co-operation in the completion of the work.

I would like to thank my parents, my family members and my beloved wife Chinmayee for being a constant source of inspiration and appreciation throughout the work. Having had the privilege of receiving such support I own responsibility for all the errors or omissions that might have occurred in the work.

Arun Kumar Nayak

1

Introduction

Dams have been built for the purpose of flood management, irrigation, hydropower production, navigation and facilitate recreation, which contribute greatly to the growth of a nation. In spite of its developmental benefits to nations, it has ignited many debates and criticisms across the world, since these projects bring significant social and human impacts, particularly in terms of displacement and raise the question of equity and justice. It is well known that the fundamental promise of large dams in the twentieth century was to provide abundant water and energy. However, it is ironical that at present we live in a thirsty world and it is likely to get thirstier. More than a billion people lack access to safe drinking water and high-yield crops and new industrial technologies often require enormous amounts of water. On the other hand, the demand of energy seems limitless. The forecasts of global fresh water and energy supplies in the decades ahead are often gloomy. However, it should not be noted at the outset that all large dams have been failures. Indeed, dams have contributed many benefits, even if these have often been distributed unevenly (Klingensmith 2007: 10–11).

The study of World Commission on Dams (2000) has noted that there has been a high degree of variability in the performance of dams, with some being plagued by numerous problems and others being relatively problem free. For example, in case of Grand Coulee (USA), the hydropower production is always higher than the prediction, where as in case of Kariba (Zimbabwe) the production keeps fluctuating from year to year. Hydropower production of Tarbela (Pakistan) has always been less than the prediction. In terms of irrigation, it has shown that only half of the predicted area of Grand Coulee is developed and it is underperforming and the same case is with Gariep dam of South Africa. But, on the other hand, Tarbela has exceeded its target by 20 per cent (World Commission on Dams 2000: 37–53). In case of flood management, none of the dams is able to check floods. For example, Tarbela dam regulates only about 16 per cent of the annual flow of the Indus River and the flood peak is reduced to 20 per cent. Similarly the same experience is being experienced in case of Glomma and Laagen basin (Norway) and Grand Coulee in Columbia basin (ibid.: 77–93).

McCully (2001), in his study, has shown that the benefits of dams are generally exaggerated. There is an important difference between a hydropower plant's capacity to generate power and its actual energy production. Over-reliance on dams lays countries open not only to the risk of wasting massive sums on unnecessary projects but also to the risks of power shortages in the case of drought, unexpectedly rapid sedimentations, miscalculated river flows or dam failure. Two dams were constructed on the Volta River – Akosombo and the smaller Kpong dam downstream, which account 88 MW of Ghana's total generating capacity of 1,160 MW. When the dams were crippled by a series of drought in 1982–83, Ghanian electricity supplies had to be rationed for the next three years. All nations in Central America rely on hydropower for well over half of their electricity. Sixty-five per cent of Guatemala's energy is provided by Chixoy dam. In 1990, the country suffered a series of blackouts because of lack of water at Chixoy. Two years later, again the county faced power shortage due to low level of water in the Chixoy reservoir. It resulted in electricity being rationed for more than a month, costing the country $2 million a day in lost industrial production. The 300 MW turbines of El Cajon dam represent 70 per cent of installed capacity of Honduras. Low rainfall through the early 1990s has meant that the level of El Cajon reservoir has steadily fallen. By mid-1994, the dam was able to generate at only half capacity at best and Honduran electricity consumers were suffering blackouts of up to 14 hours a day (McCully 2001: 133–140).

There is a great illusion that dams control floods. The US Army Corps of Engineers has spent over $25 billion on 500 dams and 16,000 kilometres of embankments in its war against floods. RuRec, the TVA and other federal, state and local bodies have also spent billions more on dams for flood management. However, the cost of flood damage in the US has more than doubled since 1937, when the first federal Flood Control Act was passed. The number of people killed in floods each year has remained roughly the same. Flood damage to property in the US in the first half of the 1990s averaged around $ 3 billion a year. This pattern of rising spending on flood controls being accompanied by rising flood damage is seen around the world. India too spent nearly a billion dollars on embankments and river channelization between 1953 and 1980, and many millions on dams, yet both area of cropland affected and the cost of damage from flooding rose dramatically over this period (ibid.: 146–148).

There are number of reasons of flood damages, such as deforestation, degradation and urbanization of watersheds and climate change, etc. Probably the most important factor behind the spiralling costs of floods around the world, however, is that dams and embankments induce a false sense of security. Dams function as structural control only eliminates 'normal' annual floods and it sometimes worsens the severity of extreme floods. A reservoir with sufficient reservoir capacity can help alleviate flood downstream by storing some or all of the excess flow after heavy rains. However, in spite of sufficient reservoir capacity of various multi-purpose dams, they fail to control floods due to financial and political pressure to keep the reservoir level high in order to maximize hydropower generations and water supply to its intended stakeholders. Numerous cases have been recorded of floods which have been made worse because of dam operations. Many deadly floods have been blamed on emergency releases of water from Indian dams. In 1978, nearly 65,000 people in Punjab were made homeless by floods exacerbated by forced discharges from Bhakra dam (ibid.).

It is being claimed that dams make permanent flood in the reservoir to control temporary flood in the downstream. The Columbia dam, for example inundated 11,130 hectares of prime farmland to provide flood protection for less than a third as many hectares in the downstream. While the original project documents discussed the benefit of flood control, however, no mention was made of the multi-million dollar loss of farm production and related business to permanent flooding by the reservoir. The massive Three George dam of China promised to save 10 million people from the ravage of floods and on the other hand it victimized 1.3 million people of permanent flooding of their homes in reservoir zone. Even if the full emergency storage capacity were utilized, the flood control is marginally normal (ibid.).

Only a tiny fraction of the water impounded by large dams is supplied for use by households and businesses. It has shown that very few large dams have water supply component and it is generally a minor purpose of the project. Although there is no global data available on the different sources of municipal water, it is likely that most of the world's municipal water supplies come from groundwater. Even in the heavily dammed US, more than half of the population are supplied from groundwater. In Europe, 65 per cent of the people are dependent on groundwater. Estimation has been made by the World Health Organization (WHO) that 1 billion people do not have access to a decent source of water. Out of those 855 million live in rural areas. The hugely expensive, sophisticated and energy intensive networks of pipes, pumps and treatment facilities that would be needed to provide drinking water to dispersed rural populations from large reservoir means that big dams are usually not even an option worth considering for rural water supply in developing countries. It can be done only through community managed schemes relying on local wells, springs, streams and rivers, etc. which will be easily maintained technologies (ibid.: 148–151).

It too has become ironical that dams dewater and lower the groundwater tables below the dams (floodplains). Perhaps the worst example is India's Farakka Barrage, which diverts water from the Ganges to the port of Calcutta and it has reduced drinking water availability for 40 million people downstream in Bangladesh. Sometimes, claims are made much more than its real potential of dams to provide water supply. In case of Gujarat in India, the Sardar Sarovar project documents claimed in 1983 that 28 million people would get drinking water in the drought-stricken state. In the year 1989, the claim was extended to 32.5 million without any further explanations and in the year 1992 it was again swollen to 40 million. However, in the following year the number fell back to 25 million (ibid.).

Proposal for multi-purpose dam projects frequently add navigation as one of their list of benefits.

However, it has not become successful as it claims. It put the traditional river user in the disadvantaged position, when their boats are not suitable for negotiating with the reservoir where drowned tree trunks and submerged buildings and villages make obstacles. The backwater sedimentation of reservoirs can also hinder navigation in their upper reaches and require expensive dredging. Downstream, the dewatering of rivers has an obvious effect upon those who depend on boats for their livelihood. The most striking illustration was made earlier of Farakka Barrage. In China, according to the World Bank, the length of navigable waterways fell from 1,70,000 kilometres in 1960 to 1,09,000 kilometres in 1984 'in part due to building of dams' (ibid.: 151–153).

Many of the proposals for the major 'multi-purpose' dams were built in Africa and claimed navigation as a benefit. Taking the case of Kanaji dam in the 1960s of Nigeria, it was fitted with a huge lock served by a six kilometre long access canal. The 49 metre lift lock, one of the highest in the world, can hold 5,000 tonne barges at a time. As of 1994, almost three decades after the lock was installed, no barge had ever used it. One of the claimed purposes of Manantali dam was to regulate the Senegal River to allow barges to carry bauxite and iron ore across 900 kilometre from Kayes in Mali to the sea. However, the dredging and construction work needed to make the river navigable and it would cost more than $400 million, which is more than the costs of Mnantali dam and unlikely ever to be done (ibid.).

In many cases, the navigation benefits only go to large scale multinational corporations at the expense of public taxes. The Army Corps of Engineers built 29 'lock and dam' projects on the Upper Mississippi between 1914 and 1950, turning 800 kilometres of river into a series of slackwater navigation pools. It costs the federal government around $12 billion dollars in 1990. However, the massive subsidy has gone mainly to benefit the 20 corporations, which own four-fifth of all barges in the US. The locks and dams are now starting to deteriorate and will need expensive refurbishment at the taxpayer's expense (ibid.).

Irrigation is claimed as one of the most important benefits of dam. However, it doesn't fulfil as it claims. Reservoirs and canals consume large amounts of prime quality of farmland. In some cases, the amount of cropland consumed is not appreciable less than the new land irrigated and in few cases it is substantially more. Bakolori dam irrigates at maximum 44,000 hectares, but it submerged 12,000 hectares of land and wiped out cropping on further 11,000 hectares downstream by stopping the annual flood. One of the biggest irrigation fiascos in India is Bargi dam on the Narmada, which submerged nearly 81,000 hectares of farmland and forest to irrigate a projected area of 4,44,000 hectares. Although the dam was completed in 1986, seven years later only 12,000 hectares were receiving irrigation water (three per cent of the planned area) (ibid.: 164–167).

The massive expansion of irrigation and its poor management of canal irrigation have been resulting in salinization and waterlogging on an unprecedented scale. An estimation made by Centre for Resource and Environmental Studies (CRES) at the Australian National University indicates that around 45.4 million hectares, one-fifth of the world's irrigated land are 'salt affected'. Around 2 to 3 million hectares become so affected by salinity each year that they have to be abandoned. Salinization of irrigated soils worldwide is costing some $11.4 billion a year in lost income due to reduced yield. One of the few comprehensive studies of the farm level effects of irrigation-induced salinity was done on the World Bank-funded Sarda Sahayak Canal irrigation project in Uttar Pradesh, India. The study found that wheat and rice yields were cut in half on saline and alkaline soils, and that the incomes of farmers in affected areas were slashed to as little as 3 per cent of those of farmers cultivating healthy soils (ibid.: 168–169).

Canal irrigation has played a central role in the 'green revolution' in Asia, especially in northern India and Pakistan.

But, critics argue that it promotes capitalism, using highly capital intensive technologies and its productions are meant for export rather than local consumptions. The absence of land reform and sustainable agricultural policies only benefits the rich farmers and deprive marginal farmers. The introduction of green revolution now made millions of Asians dependant on high-yielding variety crops and the programme is in trouble. The study made by International Rice Research Institute (IRRI), Philippines, has shown that the high-yielding varieties which were producing 10 tonnes per hectare in 1966, has now reduced to less that 7 tonnes per hectare. IRRI scientists are not yet sure about the trend of declining production of rice, but they believe that the overall reason is the degradation of paddy soils due to the switch from seasonally flooded rice cultivation to perennial irrigation with two to three crops growing annually. The flooded rice fields may not be resilient to intensive cropping which is killing important microbes of soil and causing the reduction of productions. The replacement of traditional manures with modern fertilizers may also be depleting essential trace elements in the soil such as zinc and sulphur (ibid.: 171–172).

Large modern irrigation systems also have far-reaching effect on the lives of farmers. The central administration of canals by government bureaucrats frequently intrude the decisions of farmer community and dictate their will. It is the irrigation bureaucrats, who decide the release of water, how to give, whom to give and when to give and also decides the kind of crops to be cultivated and its necessary essentials of plants, fertilizers and pesticides etc. The study of Muneera Salem-Murdock of the Institute for Development Anthropology tells how the Sudanese state owned corporation, which runs the huge New Halfa irrigation scheme, forces tenants to grow cotton despite the fact that the crop's production costs almost invariable exceed returns to the tenants, pushing them into debt (ibid.: 172–173).

Promotion of irrigation based agricultural method gradually leads to the accumulation of large tracts of land in the hand of single proprietors and it consequently dispossesses the small landholders. Iran's dam, Dez, was supposed to irrigate 80,000 hectares for the benefit of small farmers; however, after its completion in 1960s, the Shah and his advisors decided that the irrigation water would be better used by foreign agribusiness corporations producing for exports. The 'farmers' who were awarded 16,000 hectares of land, but in actual irrigation was received to Mitsui, Chase Manhatten, Bank of America, Shell, John Decree & Co. and the Transworld Agricultural Development Corporations.

The same story was narrated in Rajasthan, India. Hundreds of thousands of landless and poor farmers were given plots of land as the first phase of the huge Indira Gandhi Canal advanced across the Rajasthan desert in the 1960s and early 1970s. However, two-fifth of the settlers had lost their land by 1989. Government apathy towards farmers and absence of loan scheme with low interest rate forced farmers to take loan from private moneylenders at interest rates up to 50 per cent. As a consequence, most settlers have become deeply indebted and many became bonded labourers and now their lands are owned by moneylenders (ibid.: 173–176).

There is an assumption that there is a direct relationship between increasing crop production and reducing malnutrition. However, it is not completely true. Many of the countries are self-sufficient or there is surplus production in both the developed and developing countries, but on the other hand, major chunk of the people still live in poverty. Even in the US, California is major producer of irrigated crops, one in six residents suffer from chronic hunger. In South Asia, countries like India, Pakistan and Sri Lanka have their surplus productions, however, hundreds of millions go hungry everyday in these countries. Today, the problem is not about food production, but the problem is who produces the food and who has power to obtain it. Further, the irrigation scheme generally uses to grow expensive crops for the purpose of exports and thus it marginalizes the poor (ibid.: 177–179).

In India, wheat and rice were grown on two-third of the area newly irrigated between 1960 and 1983 and on the other hand the total amount of land under traditional cheap subsistence food crops (pulses, coarse cereals, millets, sorghum etc) fell over the same period. Although nutritionists believe that traditional diet is superior to a wheat-based diet, there is little official support for traditional subsistence crops, compared to the hundreds of billions of rupees spent on irrigation infrastructure and the promotion of wheat and rice growing. Before the Indira Gandhi Canal was made, 1,35,000 hectares acres of land in the Thar Desert were planted to coarse grains, including the traditional grain for bread, *bajra*. By 1990, however, *bajra* growing had been almost eliminated and the desert now under perennial canal irrigation, was planted with 1,32,000 hectares of cotton and 1,23,000 hectares of wheat, as well as tens of thousands of hectares of mustard and groundnut, chickpeas, sugarcane and rice (ibid.).

Modern canal irrigation has a woeful record of poor design and construction, mismanagement and corruption. Water supplies through canal irrigation to farmers are frequently erratic and unreliable as compared to groundwater irrigation. The Bakolori project of Nigeria yields only 69 per cent of groundnuts, 56 per cent of rice and 17 per cent of cotton, predicted by agricultural economists. In 1971, the Nigerian Department of Agriculture estimated that Nigeria would have 3,20,000 hectares under large scale irrigation projects by 1982; yet at the end of 1980, less than 31,000 hectares were actually irrigated by modern scheme. In India, the Public Accounts Committee reported in 1983 that since Independence not a single large irrigation project in the country has been completed on time and within stated cost estimates. Even when projects are reported as being completed they are still likely to irrigate much less land than originally planned. Over half of the irrigation potential created on many schemes in South Asia has never received any water (ibid.: 179–182).

The modern scheme of canal irrigation and its water supply distribution is invariably skewed. It distributes unequal amount of water in the Head Reach Area (HRA), Middle Reach Area (MRA) and Tail End Area (TEA), leading to uneven development. Generally, HRA gets more water than TEA and sometimes they do not get any water at all. Studies on Gal Oya project in Sri Lanka showed that TEA rice yields were only just over half of those of farmers of HRA. In India, the Sardar Sarovar project promised to bring irrigation in the most drought prone areas of Saurastra and Kutch by constructing massive canal network. However, the TEA is facing chronic water shortages. Irrigation inefficiency (measured as the percentage of water actually used for crops growth relative to the total amount of water delivered by the irrigation system) averages only around 40 per cent worldwide. In India, the World Bank notes, irrigation efficiency is often assumed in project documents at 60 per cent, where as in real life most Indian schemes probably have an efficiency of 20 to 35 per cent (ibid.).

The development benefits of dam are not only exaggerated but also its socio-economic impacts are underestimated and ignored. Large scale displacement has been justified by invoking the utilitarian logic that 'few people have to sacrifice for the national interest or goods'. However, there is no data of dam induced displacement. Not surprisingly, given their size, population densities and number of dams, India and China have displaced and are displacing more people than any other nation. Indian Social Institute, New Delhi, has estimated that dams are single largest cause of displacement accounting for about 75 to 80 per cent (40 million) of the total displacement in India. Among them, 40 per cent of the displaced are the Scheduled Tribes (STs) and 20 per cent belong to the Scheduled Castes (SCs) (Patwardhan 1999: 1–12). As per the government records, 75 per cent of those displaced are still not resettled and rehabilitated. It's a cruel joke that after 60 years of Independence, the Government of India did not wish to promulgate a National Policy on Resettlement and Rehabilitation for the displaced people (Fernandes and Paranjpya 1997: 1–34).

There are number of studies on dams and involuntary displacement and these studies have shown that construction of dams force people into new physical setting, which are alien social worlds, functioning on unfamiliar lines, and they enter at a structural disadvantage, having little educational, cultural or financial capital. The most outstanding study on involuntary displacement and its socio-economic impacts is done by Cernea, popularly known as 'Impoverishment Risks and Reconstruction Model'. He says that displacement leads to the social exclusion of certain social groups of people. It culminates in physical exclusion from a geographic territory and economic and social exclusion from a set of functioning social networks. Thus, affected people face a broad range of impoverishment risks that includes landlessness, joblessness, homelessness, marginalization, food insecurity, increased morbidity, loss of common resources and social disarticulation that result in a loss of socio-cultural resilience. Among the oustees, most of them are aboriginals or tribals. This inevitably means a transfer of power over rivers as resources: away from local control and customary law to bureaucracies and state regulation. Even rehabilitation efforts are made for the displaced by governments, it varies widely and it occurs in the context of social relationship of power (Cernea 2000: 3663–3666; Cernea 1995: 266–267).

By the end of twentieth century, the world's governments had built more than 45 thousand large dams and majorities were constructed after 1945. There are dams on approximately 60 per cent of the world's river and on nearly every large river. It displaced somewhere between 40 to 80 million people from their homes and lands. Dams too have failed to bring development benefits as compared to its claim and its huge socio-economic and environmental impacts raise number of questions that what factors compelled nations to introduce and execute such mega projects worldwide. The answer is given by Klingensmith (2007) in his study that dams are artifacts of 'utilitarian public policy', connected to certain socio-economic and political situations in general and invoking the idea of nationalism in India in particular (Klingensmith 2007).

In this context, this book has described dams and development in general and in India in particular with special reference to Hirakud dam. It has explained the broader issue of development, displacement and rehabilitation consequent to the construction of Hirakud dam. The study aimed at (a) assessing the benefits of the project as projected and (b) nature and extent of displacement and its socio-economic impacts (c) resettlement and rehabilitation facilitated to the oustees in order to understand the dispensation of justice (as conceived by Rawl) (d) how the dam was projected as the project of nationalism and justified as secular temple (e) and how the protests and movements against the dam were suppressed in the name of nationalism.

The study is based on both primary and secondary data. The primary sources consist of government documents, records and reports and the secondary sources are consisting of books, articles and periodicals. So far as displacement and its socio-economic impacts are concerned, information is collected through survey and focused group discussions with oustees. The study in its survey has taken one rehabilitation camp as sample out of its 18 rehabilitation camps, namely Kadalipal. The rationale behind selecting this rehabilitation camp was, albeit, purposive in nature, where maximum numbers of resettled families were left out. Therefore, purposive sample has been used in the study, given my time and resource constraints. Besides the rehabilitation camp, I also randomly selected one host resettled village (Jammal), where oustees have resettled by themselves with their own effort voluntarily without any governmental assistance. In this study, simple random sample has been used. The survey was conducted and evidence obtained from oustees by holding group discussions using unstructured questionnaires in both the case studies. Besides, I also consulted many local NGOs, activists, academicians and many prominent persons of that region in order to gather information for the fulfillment of the study.

The book has eight chapters. The first chapter, 'Introduction', has discussed the contemporary debates on dams and introduced all the chapters of the book.

The second chapter, 'Dynamics of Development', discusses the rationale of dam construction and its proliferation across the world, its contribution to development and its social, economic, ecological and various environmental effects. It also describes resettlement and rehabilitation policies of various countries, both in terms of land based and non-land based strategies. It too analyses how dams create uneven development, which leads to protests, movements and conflicts in the society, and how it raises the question of equity and justice. It describes how large scale of forced displacement and unsuccessful resettlement scheme creates new inequalities by depriving them not only of 'income and wealth' but also of social goods in two other ways: 'liberty and opportunity' and 'social bases of respect'.

The third chapter, 'Dams and Development in India', discusses the rationale of dam construction in India, its proliferation across the nation and its contribution to development. It too has narrated the adverse side of such development (economic, ecological and various environmental impacts), which have made the important basis of protests and movements throughout the nation. In India, it has been estimated that dams are single largest cause of displacement accounting for about 75 to 80 per cent (40 million) of the total displacement in India and among them, majority are from the poor and deprived classes of the society. It is interesting to know that people who sacrifice their lands, sources of livelihoods and various valuable assets for the sake of national interest, the state had no provision of resettlement and rehabilitation for those displaced. In 1993, it was for the first time that the Government of India drafted a rehabilitation policy under the pressure of both internal as well as external factors. The internal factors include: (1) the rise of protests and movements marked by a growing militancy, (2) national awareness of the problems of displacement, (3) and the demand for constitutional rights. On the other hand, the external pressures include: (1) environmental concerns raised by the Rio Conference, (2) concerns raised by Human Rights Conventions, (3) and the pressure come from funders like World Bank. This chapter vividly explains the responses made by the Government of India as well as various state governments with regard to resettlement and rehabilitation policy from time to time, to mitigate such adverse effects of development projects.

The fourth chapter, 'Dams, Controversies and Conflicts', discusses the construction of dams and how its potential benefits have become controversial and resulted in continuing of conflicts in India. India is a federal state and almost all the big rivers flows through more than one state. Therefore, construction of dams in a particular river simultaneously affects other states and it leads to controversies and conflicts; be it in the case of submergence of land, environmental degradations, large scale displacements, allocation of resources or sharing of costs and benefits. This chapter has described the nature and number of river water conflicts in India, particularly emphasizing on Narmada River dams project and it explains how it caused controversies and conflicts among states (Gujarat, Maharashtra and Madhya Pradesh). It explains that lack of consensus among them on the project led to litigations, public demonstrations and protests, resulting in delays and cost escalations, and even sometimes, bitter regional divisions.

The fifth chapter, 'Hirakud Dam and Development', discusses the rational of Hirakud dam construction on the River Mahanadi in Orissa, its contribution to development and its adverse socio-economic and environmental effects. The study has shown that the dam has submerged more lands and displaced more people than estimated in the feasibility report. The government only rehabilitated 8 per cent of the total displaced people, and replaced 14 per cent of the total submerged cultivable land. The displacement caused by the project forced the oustees to reel under various socio-economic crisis and marginalized them in various aspects of their life. It submerged large area of reserve forests and village forests, heavily impacted on the pattern of rainfall, which have been affecting the agriculture in the locality. The upcoming of numbers of industrial and mining activities in the catchment of the reservoir and their release of toxic industrial effluents into the Mahanadi have been causing harmful effects on the people of those areas, their livestock, and cultivated crops. Hence, people are being forced to adopt alternative means of livelihood. Although, the dam promised to bring development in the region, it still remains one of the most backward regions of the state. The empirical scrutiny has shown that none of the developmental objectives was fulfilled even after its 50 years of completion. In spite of bringing development in the region, now the dam has created a pattern of uneven development. There is uneven distribution of irrigation water in the Head Reach Area (HRA), Middle Reach Area (MRA), and Tail End Area (TEA), leading to uneven development in command area, which is causing conflicts among the farmers. On the other hand, the increasing diversion of water by various industries from the Hirakud reservoir especially after the economic reforms have been causing scarcity of water in the TEA and, therefore, conflict has also been arisen between farmers and industries over the water of Hirakud reservoir.

The sixth chapter, 'Idea of Multi-Purpose River Valley Development (MPRVD)', discusses the rational of the emergence of Multi-Purpose River Valley Development (MPRVD) model of the US and its proliferation across the world and has examined that, to what extent; the replication of MPRVD in India was relevant to execute it with special reference to Mahanadi Multi-Purpose River Valley Development plan (Hirakud dam). It describes that the genesis of TVA in the US was connected to the particular socio-economic and political situations of 1930s. This period was marked by severe economic recessions and TVA was established in 1933 in order to immediate stimulation of employment in the region. The era was also marked by the rapid decline of the European colonial powers and the emergence of the United States of America as the new hub of capitalist dynamism. Many governments, including India, adopted MPRVD immediately during their post-colonial phase with the help of American experts for its construction to speed up their economic development. The fundamental promise of large dams in the twentieth century was to provide abundant water and energy. However, at present we live in a thirsty world and it is likely to get thirstier. On the other hand, the demand of energy seems limitless. The forecasts of global fresh water and energy supplies in the decades ahead are often gloomy. These projects remain far away to fulfil such intended objectives. The empirical scrutiny has shown that Hirakud dam too has failed to fulfil its promises. It is evident from the study that the model of TVA was blindly executed on the river Mahanadi without understanding the local feasibility. The integrated river basin management in India was largely promoted by many scientists, engineers and statesman, inspired by United States and the MPRVD scheme was inextricably linked to the political economy of decolonization and nationalism rather than development.

The seventh chapter, 'Protests and Movements', has described the rationales and dynamics of protests and movements against dams in general and Hirakud dam in particular. It is well known that protests and movements contribute to the legitimacy of democratic governance, as they significantly enhance the interactive space between the rulers and ruled. Protests and movements in a democracy are more extensive and less deadly than those under authoritarian rule. Contextualizing the case of Hirakud dam, there was large-scale opposition from the local population as soon as it was decided to build dam at Hirakud. Although the nature of the domestic opposition to the Hirakud dam was strikingly similar to domestic anti-dam struggle that emerged in the greater numbers in India during the 1970s, it failed to prosper due to the absence of NGOs, transnational allied advocacy networks, legitimized global norms on human rights, indigenous peoples and the environmental lobbies. The domestic resistance to Hirakud dam was politically too weak to influence the institution of India's democratic regime to either halt or even reform the Hirakud project. Besides the failure to attract larger alliances, the resistance to Hirakud dam was overshadowed by the nationalistic rhetoric of nation building in these early years of post-Independence. Even if India is a democratic country, it acted like authoritarian state and suppressed the democratic voice and rights of the people. The resistance was quite justified on the ground that none of the objectives has been fulfilled even after its 50 years of completion. On the other, it led to large scale socio-economic and environmental consequences. Even, the proposition to construct a dam at Hirakud was opposed by an engineer (M.G. Rangaya) in his report; the government did not pay any attention to his suggestions and now the dam is facing the exact situation, what he has predicted.

The last chapter, 'Conclusion', has raised number of questions and made the conclusion of the study and recommended certain suggestions with regard to the relevance and need of dam projects. It asks that should large dams be built? The unfortunate answer is yes. Large dams remain a necessary development option for providing water and energy resources to populations in developing countries that are in crisis. Taking the case of India, its growing population and rising need for food and water present a difficult situation. Further, it is compounded by dwindling monsoons that concentrate rainfall within a relatively short time period and in such situation, it is hard to see an alternative in some large storage reservoirs. But dams should only be constructed after a 'best practices' options assessment process that gives sufficient emphasis to environmental and social issues and only where adequate policies exists and implemented and where project authorities, contractors and consultants have a legal responsibility to follow whatever contractual conditionalities are necessary to implement the project as intended. Besides, dams should not be constructed in the most favourable zone of matrix that gives extensive weightage to only high technical advantages. The process of dam building should be based on a through analysis of alternatives. Efforts for alternative dam site should be identified on the same river to fulfil the same objectives with least social, cultural and environmental impacts.

2

Dynamics of Development

Civilizations and Water Management

Growing population and rising levels of economic activities increase human demand for water and related services. The spree of developmental activities has brought significant use of water and its demand has been steadily increasing throughout the world. Although the amount of water resources is enough for the entire world, their distribution in time and space shows an uneven pattern. Taking these into account, emphasis is given to supplying water in sufficient quantity and quality at the right time and place, reallocating water according to certain priorities among sectors, by using surface water and groundwater resources. Among those, the engineering of dams, which provides regular water from reservoirs of dams to be used in response to the demand pattern is a vital part of civilization. Dams have played a key role in development since the third millennium BC, when the first great civilization evolved on major rivers, such as the Tigris-Euphrates, the Nile and the Indus. From these early times, dams were built for flood control, water supply, irrigation and navigation. However, construction and operation of dams received high priorities since Industrial Revolution to produce electricity. From the beginning of human history, for almost 5000 years, dams have served to ensure an adequate supply of water by storing water in times of surplus and releasing it in times of scarcity and, thus, checking floods and making a significant contribution to the efficient management of finite water resources that are unevenly distributed and subject to large seasonal fluctuations. In other words, dams can be said as a mechanism of water resource management to harmonize the natural hydrological regime with human needs for water and water related services and it is claimed as the process of development for its multi-purpose utility – flood control, hydropower production, irrigation, and navigation, water supply for industries, urban cities and drinking water for the people at large. To meet these purposes, a number of dams (especially multi-purpose river valley development) were constructed to control and regulate the natural flows of rivers. This regulation function is obviously the main reason for creating reservoirs by constructing dams (Altinbilek 2002: 9–13).

Origin and Proliferation of Multi-Purpose River Valley Development (MPRVD)

Falling water is the source of one-quarter of the world's electricity today. Despite the billions of dollar spent on nuclear power, hydropower provides a great deal of electricity on a worldwide basis. Given the myriad economic and environmental problems facing the nuclear power, hydropower maintains a substantial lead over nuclear energy. Compared with other sources of electricity (oil, coal and nuclear energy), hydropower has environmental advantages. In a world suffering from inflation and fossil fuel depletion, hydropower offers stable prices and permanence. Hydropower complexes will be producing power long after the oil wells run dry and coal fields are exhausted, if properly managed. It was during the nineteenth century that hydropower became a major source of electricity as well as of mechanical power. In 1820, the French engineer Benoit Fourneyron invented the Turbine. In 1882, it was first used to produce electricity and it steadily increased the world's electricity supply (Deudney 1981: 5–19).

At first, the global model of MPRVD was pioneered in the United States in the early 1930s with the establishment of Tennessee Valley Authority (TVA) and the construction of series of dams on the Tennessee River and its tributaries. By the later half of the 1930s, the TVA's experience and its successes was emulated by other countries across the world. In the subsequent years, engineers, agronomists, economists, regional planners and numbers of scientific and technical staffs associated with TVA were deployed and charged with the mission of replicating MPRVD in different river basins and TVA module became the icon of modernity. The era was also marked by the rapid decline of the European colonial powers and the emergence of the United States of America as the new centre for capitalist dynamism. This was one of the first comprehensive multi-purpose schemes ever deployed and dealt simultaneously with flood control, hydropower generation, navigation, irrigation, soil conservation and public health (D'Souza 2006: 184).

Referring to the development strategy after the Second World War or in the post-colonial era and economic growth as its indicator, all the countries, both developed and developing, started to grow their economy. The availability of adequate infrastructure facilities is vital for the acceleration of economic development of a country. Governments across the world have given high priority to investment in sectors such as railways, roads, power, telecommunications, ports and industries, etc. Thus, dams are the outcomes of this process and symbols of development and their multi-purpose utility – generation of electricity, irrigation, flood control and navigation – which contribute greatly to the growth of a nation (Joyce 1997: 1050–1055; Bandhyopadhyay et al. 2002: 4108). At first, this technology got widespread in the developed countries in the 1930s and also shifted increasingly to the Third World. The success of large dams in the developed countries has drawn many Third World leaders to this energy source and pronounced these as highly visible symbols of progress that facilitate the growth of heavy industry and enhance the prestige and value in developing countries (Deudney 1981: 6–19). The gospel of TVA seemed more promising abroad and since 1940s onwards, there were plans for new 'TVAs' in India, China, Palestine, Peru, Iran, Colombia and several other countries. There were plans for river development in these lands with varying degrees and it invoked as their inspiration and called for TVA personnel for their help to implement them (Klingensmith 2007: 68).

The big dam projects increasingly became more popular and contributed to the formation of various informal international big dam regimes. A group of engineers, builders and bureaucrats established a transnational professional association called as

International Commission on Large Dams (ICOLD) to collect information and exchange of knowledge about big dam building around the world in 1929. In the beginning of 1930s and 1940s, big dam building bureaucracies, including national and international agencies like RuRec or the Soviet Union's Hydrological Planning Agency and river valley organizations like the TVA, proliferated around the world. The wave of decolonization and state formation contributed to the spread of these institutions in other parts of the world and especially in the Third World. The World Bank which soon began to emerge as the premier multilateral development agency supported the big dams' buildings across the world. The growth of private dam industry and other developmental aid and credit agencies further contributed to the spread, growth and legitimacy of big dam building across the Third World. Such prominent transnational corporations are Asea Brown Boveri, Siemens, GES Alsthom, Kvaerner, Bechtel, Acres and other expanded rapidly during the 1950 and 1960s. Other multilateral organizations besides World Bank are the specialized wing of United Nations, especially the Food and Agricultural Organization, UN Development programme, Inter-American and Asian Development Bank (ADB) and bilateral aid and credit agencies like the British Overseas Development Administration (ODA), the United States Export Import Bank increasingly funded for dam buildings in developing countries (Khagram 2004: 6–7).

There was rapid increase in large dam building during the last century. In 1900, there were only 600 big dams in existence and many of them were built in Asia and Africa. It grew about 5,000 big dams by 1949 and by the end of the twentieth century over 45,000 large dams were built over 140 countries. Thus, over 90 per cent of big dams were built over the last 40 years (ibid.: 5–6). In fact, the top five dam-building countries today account for nearly 80 per cent of all large dams worldwide. China, which had only 22 dams prior to 1949, has built around 22,000 large dams, close to half of the world's total number. Other countries among the top five dam building nations include the United States with over 6,390 large dams, India with 4,000 and Spain and Japan with between 1,000 and 1,200 large dams each.

Estimates suggest that 1,700 large dams have been under construction in other parts of the world in the last five years. Of this, a total 40 per cent were reportedly being built in India (World Commission on Dams 2000: 8–10).

Dams and Development

One of the basic requirements for socio-economic development throughout the world is the availability of adequate quantities of water with the appropriate quality and an adequate supply of energy. Among all the sources of water, dams contribute significantly toward fulfilling our water supply and energy requirements, if it properly planned, designed, constructed and carried out regular maintenance. Today, dams contribute enormously towards irrigation, water supply, flood control, recreation, navigation and fish farming and other purposes in the world too.

Flood Control: Flood plain areas are ecologically important wetlands and had competitive advantage for human settlement in the past. But 'flood hazard' is a threat to life and property, health, livelihoods and other valued resources presented by a body of water, which might rise and flow over land that is not normally submerged. Flood disasters are among the world's most frequent and damaging types of disaster. During the later half of the twentieth century, floods were the most common type of geophysical disaster, generating over 30 per cent of all disasters 'between' 1945 to 1986. Globally flood disasters are about the third most harmful form of geophysical disaster in terms of loss of life. Earthquakes and tropical cyclone kill more people than any other geophysical disaster, but in the 1986–1995 period floods appear to have caused more deaths than any other geophysical disaster. The floods alone caused 55 per cent (3,67,000 people) of the deaths. The estimation of deaths caused by flooding in 1999 was 55,360 and among them 20, 000 had died only in Venezuela alone caused by mud floods and landslides. In terms of regional significance, flood related economic losses in Asia exceeded those in North America and Europe between 1987 and 1996. Thus, flood hazard management is the most essential for protecting the human being for protecting himself from such calamities (Green, Parker and Tunstall 2000: 2–4).

For centuries, societies have built levees and embankments along riverbanks to contain and control floods. The purpose has been to occupy floodplains for agricultural, urban and industrial uses and to reduce any resulting threat to lives and property. Indicators of the benefits derived from flood control include reductions in the area flooded and prevention of any consequent loss of life, social disruption, health impacts, property and economic losses. Now, about 13 per cent of all large dams in the world – in more than 75 countries – have flood management functions (World Commission on Dams 2000: 15, 58–59).

The Aswan dam is an example of a dam that controls the flow of the river to manage floods. It can store 1.5 times the average annual flow of the Nile River and has provided a high degree of protection to the lower Nile against untamed water flow of the river. The report of World Commission on Dam (WCD) has documented many dams, serving the purpose of flood control, although none of these dams were built primarily for this purpose. The Tarbela dam regulates about 16 per cent of the annual flow of the Indus River and the flood peak is reduced by 20 per cent. Similarly, Glomma and Laagen basin (Norway) and Grand Coulee in Columbia basin (USA) have seen reduced flood flows in these areas. Japan is one of the top five dam building countries in the world, and flood control is the major purpose of many dams there. It has been estimated that 50 per cent of the Japan's population lives in flood prone areas and 50 per cent of the municipalities are made victim of it. In such scenario, dams have the major role in protecting the people from floods (ibid.: 15, 58–59). However, dams are not able to control floods all the times. The most important among those causes are reducing water storage capacity of the reservoirs due to over silting or deposition of sedimentation with water flow exceeds the expectations at the time of feasibility study of dam (Adams 2000: 1–16).

Hydropower Production: Energy is the prime mover of a country's economic growth. Availability of energy with required quality of supply is not only key to sustainable development, but commercial energy also has a parallel impact and influence on the quality of service in the fields of education, health and, in fact, even food security. Hydropower is a renewable, economic, non-polluting and environmentally benign source of energy. Hydropower stations have inherent ability for instantaneous starting, stopping, load variations, etc. and help in improving reliability of power system. Hydro stations are the best choice for meeting the peak demand. The generation cost is not only inflation free but reduces with time. Hydroelectric projects have long useful life extending over 50 years and help in conserving scarce fossil fuels. They also help in opening of avenues for development of remote and backward areas. Thus, electricity generation is an important reason for building large dams in many countries either as the primary purpose or as additional functions for other purposes. Hydropower currently provides 19 per cent of total electricity supply over 150 countries. One-third of the countries in the world currently depend on hydropower for more than half of their electricity needs. Among them Canada, United States, Brazil, China and Russia account for more than half of the world's hydropower generation (World Commission on Dams 2000: 14; Milewski, Erge and Roquet 1999: 1–10; Zhang 1999: iii–iv).

There have been various studies on water across the globe, which were influenced by the new Tennessee Valley Act (TVA). The total power generation of TVA was about 30,000 k.w. in 1932 and in course of 10 to 12 years, it increased to nearly 20,00,000 k.w. It enhanced the standard of living, new industries came into being, and agriculture was improved. The doubters of yesterday became the most ardent supporters of the new order brought by the TVA. Of the 12,500 million units generated in 1946, 96 per cent was hydropower. Similar experiences were experienced in former USSR, Canada and Sweden as well. The Dnieper dam in former USSR was completed in March 1932 and it transformed the backward agrarian economy into prosperous agricultural and industrial hub. Canada affords a good example of the transformation from backward agricultural economy to industrialization, unprecedented prosperity and power. Canada lacks coal but she has immense water resources in her snow fed perennial streams. In 1945, the per capita use of electricity in Canada was the highest in the world. Sweden is another example where people have risen to great heights in industrialization and standard of living. In 1945, Swedish industry in general was almost entirely electrified and this applies also to handicrafts and home industries. By that time all the railways were electrified in Sweden. The countryside and agriculture are provided with electric light and power to a great extent (Government of India 1947: 59–65).

Looking at the hydropower performances of dams, it can be categorized into two groups. At first a number of projects that have greatly exceeded their targets and a few notable under-performers. The Cross Check Survey of World Commission on Dams (WCD) shows that higher than expected output of hydropower generation depends primarily on the extra installed capacity prior to the commissioning. For example, the Tucurui dam (Tocintans Rivers, Brazil) increased its installed capacity from 2700 to 4000 MW. Tarbela, (Indus River Basin, Pakistan) Grand Coulee (Columbia River Basin, USA), and Kariba (Zambezi River, Zimbabwe) dams have all seen subsequent installations of additional capacity that were not foreseen at feasibility. But, the increase in generation of hydropower does not always depend on the extra installed capacity rather sometimes on normal variations in weather, river flows and demand supply factors and, thus, keep fluctuating in output from year to year. Apart from this, adding additional powerhouses, upgrading existing turbines and generation equipment are the additional factors in improving performances. In case of Grand Coulee, the hydropower production is always higher than the prediction, where as in case of Kariba the production keeps fluctuating from year to year. Hydropower production of Tarbela has always been less than the prediction (World Commission on Dams 2000: 50–53).

Irrigation: Agriculture plays a greater role in the economic and social development and is seen as a vehicle for development, poverty alleviation, rural development and solution for food insecurity. It is estimated that dams irrigate 40 per cent of the total irrigated land worldwide, which contribute about 12–16 per cent of the world food production. Many countries including China, India, Egypt, Indonesia and Pakistan rely on irrigated land for more than half of their domestic production. The area of irrigated land expanded dramatically during the first years of the green revolution in the 1960s and it led to significant yields and brought down food prices. However, the scale and significance of irrigation of large dams for irrigation varies significantly from country to country. Dams supply the water for almost 100 per cent of irrigated production in Egypt – most coming from Aswan High Dam – while in Nepal and Bangladesh, dams provide only 1 per cent of the irrigation water. Four countries, China, India, the United States and Pakistan account for more than 50 per cent of the world's total irrigated area (Sanmuganathan 2000: 22–57; Gritzinger 1987: 14–15).

However, mere construction of dams or canals does not automatically ensure the lands get irrigated. A number of other factors such as water delivery, cropping patterns, intensity and yields, value of production, and net financial and economic benefits determine the success of an irrigation project. The Cross Check Survey of World Commission on Dams demonstrates that many of the dams have fallen short of physical targets, failed to recover their costs, and have been less profitable in economic terms than expected. Almost half of those have fallen short of their planned targets and poor performance is most noticeable during the early periods of the project life. The Grand Coulee, although commissioned in 1941, only half of the predicted area has been developed in the Columbia Basin Project and the same case is also repeated in case of Gariep dam on the Orange River in South Africa. However, the case of Tarbela dam is different, it has exceeded its target by 20 per cent over its 25 years.

The under-achievement of targets for irrigated area development from large dams has a number of causes, institutional failures often being the primary ones. These refer to the inadequate distribution of canals, over centralization of administration dealing with canals and inadequate attention to drainage. Moreover, over optimistic projections of cropping patterns, yields and efficiency of irrigation; the late realization that of some areas were not economically viable; and mismatch between the assumptions of the planning agency and the actual farmer behaviour are the other major factors responsible for non-fulfillment of targets (Sanmuganathan 2000: 53; World Commission on Dams 2000: 44).

Water Supplies to Urban Cities and Industries: The increasing urbanization demands effective service facilities. The most important among them are safe drinking water. Urban

Water Sector (UWS) is a zone of serious mismanagement. Typically, the large urban areas represent concentrated demands, both due to large populations and large per capita use and waste. Most urban areas have depleted, polluted or destroyed their local sources of water like rivers, lakes and tanks and in many cases even groundwater. The rainfall is generally seen as a bane rather than boon as it brings floods because the drainage systems are seriously ill designed or mismanaged.

An adequate and dependable source of water is needed both to sustain existing civilization and to support future growth. In the past and in many regions of the world today, the main sources of domestic and industrial water are groundwater or aquifers (layers of sandy gravels or rock beneath the ground surface which contain and are capable of storing water). Today, the withdrawal from many of these aquifers exceeds the natural recharge, which results in lowering of the water table. This situation is leading to depletion of ground water both in times of drought and permanently too. Of the rainfall falling on the earth, only 19 per cent falls on our land mass and a large portion ends up as runoff leaving only 2 per cent which is infiltrated to replenish our groundwater. Therefore, in order to accommodate the variations in the hydrologic cycle, dams and reservoirs are needed to store water and then provide more consistent supplies during shortages. Large urban areas depend heavily on water stored in reservoirs and it is especially critical in arid regions of the world. Water stored in reservoirs is also used for industrial needs. These needs ranges from the direct use in manufacturing to chemical and refining processes and for cooling at conventional and nuclear power plants. Managed flows from reservoirs can be used to dilute discharged substances by augmenting low river flow to maintain water quality at safe limits (International Commission on Large Dams). For example, the Western Cape region of South Africa heavily depends on six principal dams (Theewaterskloof dam, Wemmershoek dam, Steenbras dams, both Lower and Upper, Voelvlei dam, and Berg river dam), located in the Cape Fold Mountains to the east of Cape Town. These dams provide 63 per cent of water supply to Cape Town, meant for domestic and industrial purposes. These too provide 5 per cent of its supply to smaller towns for the same purposes and rest 32 per cent to agriculture for irrigation to different areas (Wikipedia, Western Cape Water Supply System). Likewise, in China, Capitol Steel Mill at Beijing takes 2,46,000 litres (65,000 gallons) to produce 1 ton of steel. The Almendra dam of Spain also provides huge amount of water to pulp and paper mill for their industrial uses (International Commission on Large Dams).

Navigation and Recreation: It is generally known that carriage by water is cheaper than any other form of transport. In case of heavy goods and cargoes, water transport is more favourable than the transport by railway. The economy of a country depends primarily on its internal as well as external trade, which increase the standard of living of the people. Comparing the advantages of inland navigation with highway and rail, the former has large load carrying capacity and the ability to handle cargo with large dimensions. Inland navigation has the lowest fuel consumption and least amount of pollution of the bulk transportation modes. It contributes to reducing the emission of greenhouse gases and, thus, to mitigating the worldwide effects of global warming. Thus, dams ensuring enhanced inland navigation provide a vital role in realizing regional and national economic benefits. Additionally, it too reduces erosion and stabilizes groundwater levels throughout the system (International Commission on Large Dams).

The substantial increase in India's future sea trade can be estimated by drawing comparison between the highly developed and industrialized countries. For example, the total volume of India's and coastal sea-borne trade during 1939 was approximately equivalent to 0.05 tonnes per head of the population, whereas in the US it was 3 tonnes per head, and for the UK about 4 tonnes per head. The two largest ports (Calcutta and Bombay) in India handled about 15 million tonnes of goods in 1939, whereas London and Liverpool the two largest ports in the UK handled during the same year a total tonnage of over 63 millions. The total tonnage of goods handled in 1939 by all Indian major and minor ports amounted to 21 millions, which was less than the total handled by Liverpool alone during the same period. It was remembered that the population of India was about 10 times that of the UK and the total sea-borne trade in goods was less than one-half of that handled by London alone. Therefore, dam as a technology was adopted by various countries across the globe to develop navigation by taming rivers to accelerate their trade potentials (Government of India 1947: 67–70).

However, only few rivers in the world are readily navigable, largely due to natural obstacles, such as shallow stretches of the river bed or too a high velocity of the flowing water, that have to be overcome or eliminated by the construction of locks, dams and control structures. The attractiveness of reservoirs for recreation is often adds a significant benefit. This is very significant in areas where natural surface water is scarce or non-existent. Recreational benefits associated with lakes, such as boating, swimming, fishing, birds-watching and nature walks. These are taken into account early at the planning stage, which makes project a balanced project and it enhances recreational opportunities (International Commission on Large Dams).

Development and its Discontents

While many have benefited from the services dams provide, but their construction has led to many significant social and human impacts, particularly in terms of environmental costs and human displacement. Compulsory displacement that occurs for development reasons, embody a perverse and intrinsic contradiction in the context of development. They raise major ethical questions, because they reflect an inequitable distribution of development's benefits and losses. Nevertheless, the involuntary displacements caused by such programmes create major impositions on some population segments. It restricts population rights by state-power intervention. This raises major issues of social justice and equity (Cernea 2000: 3659).

Environmental Costs: In fact, the water flow regulations by dams provide benefits for many segments of society and worldwide, many millions of people have benefited from their construction. But on the other hand, conversely, many people have suffered and much environmental damage occurs as a consequence of dam construction. It is estimated that, now, there are more than 40,000 large dams throughout the world, with an aggregate storage capacity of about 6,000 km^3 and this represents a seven-fold increase in the standing supply of natural river water. It is estimated that about two-third of the freshwater flowing to the oceans is controlled by dams. However, for most of the world's existing stock of large dams, environmental issues have played little part in their design and the specification of operating rules. Most dams have been constructed with the emphasis on maximizing the economic use of water, with little or no understanding of the long-term consequences of alterations to flow volumes, flow patterns and water quality.

This marked a turning point in modern thinking and raised several important questions. The publication of the Brundtland Report, 'Our Common Future', in the year 1987 put forward two interrelated concepts. The first is 'sustainable development,' which has been defined as 'development that meets the needs of the present without compromising the ability of future generations to meet their own needs'. The second concept is that the lives of people and the environment are profoundly inter-linked. This has led to an alternative definition of sustainable development, i.e. 'improving the quality of human life while living within the carrying capacity of supporting ecosystems' (McCartney, Sullivan and Acreman).

Institutional and Political Context of Displacement: The first and most important factor that displaces people is the prevailing 'Land Acquisition Act' (LAA) in different countries. This empowers states to enjoy unlimited power over land to acquire it within its territory for 'public purpose', which cannot be legally challenged by a person or community. The doctrine of 'eminent domain'[1] in India enjoys the power for the acquisition of land (Baxi 1989: 164–171; Hemadri 1999: xxxii–xxxiv; Bartolome et al. 2000: 9; Ramanathan 1995; Sharma 2003: 907–910; Dankalmair and Bartolome 1999: 1). The same kind of law is also prevailing in Ghana. The Ghanaian State Lands Act of 1962 empowers the state to acquire lands for the sake of national interest (Bartolome et al. 2000: 9). Similarly, the Public Interest Appropriation Law is used in Argentina to acquire land with just monetary compensation without any provision of resettlement or other kind of compensations and mitigation measures.

Although Article 41 of the new Argentinean Constitution (1994) establishes the need to protect the environment and the obligations of mitigating the socio-economic impacts, the article, yet has not developed into specific norms and it is open to diverse interpretations (Dankalmair and Bartolome 1999: 1).

Second, 'international institutions', especially those who fund for the buildings of dams worldwide, are another cause of accelerating the number of displacement.

- Multilateral banks: such as the World Bank, Asian Development Bank, Inter-American Development Bank
- Bilateral aid agencies: CIDA (Canada), SIDA/BITS (Sweden), USAID (United States), JICA/OECF (Japan), DFID/CDC (United Kingdom) and NORAD (Norway)
- Export credit agencies: Exim Bank (United States), Jexim (Japan)
- UN agencies: UNDP

Among them, World Bank, Asian Development Bank and Inter-American Development Bank are most important international institutions fund for dams building worldwide (Bartolome et al. 2000: 9–10).

Third, 'privatization and the market forces' in the post-liberalization era is another critical institutional factor extending the number of displacement. The spree of market forces and private capital and their large scale infrastructural projects have been displacing people by acquiring their private lands for public purpose. Here, the most significant questions centre on the relationship between private capital, public interest and social costs and it is well illustrated in case of Mexico. The Huites multi-purpose dam on Sinaloa, Sonora and Chihuahua state boundaries in northwest Mexico is first privately financed mega dam in the Mexican history. This project was begun in tandem with the 1992 reforms to Electrical Sector and Water Resources Laws permitting private capital joint ventures for the first time since the revolution of 1910. The Huites project, under the provision of the Central Water Commission (CWA), called for selling of water to farmers on 70,000 hectares of the arid coastal plain of Sinaloa and Sonora states. The important issue here is the resettlement of the indigenous Mayo people, those who were displaced from the project.

There is neither effective legal framework guaranteeing the rights of the Mayo, nor the accountability of the project officials under the new rules of private sector investments. The cases of Huites imply the absence of institutional, legal and policy framework for the displaced. This may well be the emerging context in Mexico and elsewhere in Latin America: a state legacy of impunity in the process of rampant privatization, ignoring the rights, human and environmental concerns, those affected by infrastructure projects (Robinson 1999: 1–9).

Social Context of Displacement: Large dams have significantly altered many of the world's river basins with involuntary impacts on the 'livelihoods' and socio-economic foundation of millions of people. The livelihood impacts caused by dams on people are particularly devastating in Asia, Africa and Latin America. The Cross Check Survey of WCD shows that physical displacement has taken place in 68 of the 123 dams (56%). Of the dams in the sample 52 out of 68 dams are in Latin America, Asia and Sub-Saharan Africa, the most populated regions of the world. The scale and extent of impacts will vary depending on location, size and other dam characteristics such as inundated area and population density in the river basin. The scale and extent of impacts will vary depending on location, size and other characteristics, such as inundated area and population density in the river basin. The number of people deriving their livelihood directly from the river and its ecosystem, and the overall population density in the river basin, gives an indication of the potential impacts (World Commission on Dams 2000: 102–104; Sharma 2003: 907–908; Bharati and Rao 1999: 1374–1375). Therefore, China and India, the two most populous countries in the world, account for the largest number of displaced people, and they share around 57 per cent of the world's large dams. In China, large dams are estimated to have displaced about 27 per cent of all people displaced by development projects (the total includes people displaced by urban expansion, roads and bridges), while in India, it figures 77 per cent (this total excludes people displaced by urban development). Although China is highly populated country and has constructed large number of dams in comparison to India, the latter exceeds the number of displaced than the former. Here the population density inhibited in the river basins is the main factor determining the large number of displacements. The population density in India figures 374 persons per square kilometre, whereas, it is 224 persons in China (World Commission on Dams 2000: 16–17).

The extensive study by Michael M. Cernea on involuntary displacement and its socio-economic impacts is popularly known as 'Impoverishment Risks and Reconstruction Model'. His study shows that displacement leads to the social exclusion of certain social groups of people. It culminates in physical exclusion from a geographic territory and economic and social exclusion from a set of functioning social networks. Thus, affected people face a broad range of impoverishment risks that includes landlessness, joblessness, homelessness, marginalization, food insecurity, increased morbidity, loss of common resources and social disarticulation that result in a loss of socio-cultural resilience (Cernea 2000: 3663–3666; Cernea 1995: 266–267).

Among the displaced, 'indigenous people' or the 'tribals' (ethnic minorities) are the worst victims in the whole episode of displacement. In Philippines, in case of almost all the large dams, those that have been built or proposed, the most vulnerable are the indigenous people, who constitute 6 to 7 million. Similarly in India, 40 per cent of those displaced by development projects are indigenous people, who account for just 8 per cent of the total population of the country (World Commission on Dams 2000: 110).

The sources of livelihood (fodder, fuel wood, fibre and fruits) of indigenous people are closely associated to the forestland; they rely on number of activities, including hunting, fishing, gathering and shifting cultivation. They also cultivate their ethnic agricultural products in the forests land, suited to their climatic conditions. In comparison to other section of the society, they also tend to be more vulnerable, as they often lack formal rights to the area, on which they depend on for their subsistence. The absence of legal right over their traditional land, therefore, deprived them from compensations and any form of resettlement or rehabilitation. It became too complex for them to sustain, when their livelihood pattern shifts from 'forest economy' to non forest economy (Shylendra 2002: 3289–3290; Mahapatra 1991: 272–273; Jena 1998: 822). The indigenous people living in the rural areas, often do not have formal education and may face language barrier, cultural obstacles and racial prejudice as well as lack financial resources and social and political contacts to influence the decisions that are made in the national and regional capitals (Inter-American Development Bank 1998: 26–27). They remain deprived of justice due to negligence and incapacity to fight for their rights. Structural inequalities, cultural dissonance, discrimination and economic and political marginalization are the main factors responsible for their marginalization (World Commission on Dams 2000: 110).

There are many cases of development induced landlessness and marginalized indigenous people across the world. The vulnerability of indigenous people in case of both developed and developing countries are similar. The Missouri River Basin of the United States during the 1950s and 1960s submerged 1,42,000 hectares of the best land, belonging to the indigenous people, including number of burial and other sacred sites (ibid.: 111). Similarly, the Mandira dam constructed on the river Sankh for the purpose of supplying water to the Rourkela Steel Plant Project in India snatched away huge area of tribal land and displaced about 1,00,000 people, without any provision of compensations and rehabilitation (Nayak 2000: 79–108). It can be stated that, the rights of the indigenous people and the ethnic minorities are poorly defined or enshrined in the national legal framework, and lack effective protection (World Commission on Dams 2000: 111).

Environmental changes and social disruptions caused by large dams and associated infrastructure developments can have significant adverse impact on 'human health' for displaced communities. The issue of equity in terms of pre-existing nutritional and health conditions of the population and the capacity to resist new health problems is at the root of the adverse health impacts of dams (World Commission on Dams 2000: 118). The health issues associated with the displaced can be both communicable as well as non-communicable diseases. There is regional variation in the prevalence rate of health hazards and it depends primarily on ecological factors; such as the presence of insect vectors, types of vegetation and rainfall (World Health Organization 1999: 4–11). Communicable diseases such as yellow fever, rift valley fever, onchocerciasis, trypanosomiasis are not found in Asia but has wide distribution in Africa and tropical America. Communicable diseases such as Japanese enchapelitis and dengue fever are found in Asia region. Opistorchiasis is an example of a parasitic disease that is most common in South East Asia. But it is restricted to the belt of China, Philippines, a valley in Sulawesi and small section of the Mekong River. It sometimes leads to psychological trauma accompanied by the outbreak of relocation related illness, particularly parasitic and vector-born diseases; such as malaria. Unsafe water supply and improvised sewage system increases vulnerability to epidemics and chronic diarrhea, dysentery, etc. The weakest segments of the demographic spectrum – infants, children and the elderly – are affected most strongly (Cernea 2000: 3665). The WCD shows that at least 50 people died in the Lisitu area on the strong emotional response of the Gwembe Valley Tonga to their involuntary resettlement caused by the Kariba dam in Zimbabwe. Many of them also died of dysentery and measles that occurred in the earlier years of their displacement (World Commission on Dams 2000: 119).

'Gender' is a vital element while understanding the impacts of involuntary displacement. It not only results in terms of losses and destructions of goods and property, but also alters the people's lives and their social fabric. The displaced coming to the new environment loses their former support structure and it affects the social roles and responsibilities of men and women. Women have major role in contributing their household livelihoods and their dislocation results in loss of their livelihoods (Gururaja 2000: 13) and such situations promote them for alcoholism, prostitution, gambling, etc. and it affects the lives and status of women. They also sometimes face physical and sexual violence in a newly relocated place (Asian Development Bank 2003: 1–6). The Cross Check Survey of WCD found that women suffered more than men in case of Tarbela dam, separating from their ancestral land and their relations from water, forests and natural resources. Similar experiences were also found in case of Pak Mun dam (Thailand) and Akosombo dam (Ghana). Increased alcoholism and domestic violence were rampant in the resettlement sites of Sardar Sarovar dam (India) and Kariba dam (Zimbabwe) (World Commission on Dams 2000: 115).

In the post-displacement scenario, women generally are denied compensations, property rights and access to land. In a patriarchal society, land and property are passed to male heir; women only enjoy the user rights over land and forest, which is mostly prevalent in Asia. Thus, the absences of legal title of women exclude them from compensations for their lost lands and properties. The prevailing unfavourable deal also causes more victims to the widows, divorced women, elderly and unmarried (Asian Development Bank 2003: 1–6). In case of Mahewali dam (Sri-Lanka), Sardar Sarovar Project (India), only the sons were nominated for land compensations (World Commission on Dams 2000: 114).

In the process of involuntary displacement, conflicts also arise in many cases between the displaced people and 'host communities'. It puts heavy pressure on the resources of host communities and competitions take place over the sources of livelihood and employment. The inclusion of non-homogeneous group on the basis of caste, religion, ethnicity and community creates cultural clashes, social tensions, political problems and demographic imbalances. Sometimes these problems become so severe and lead to caste conflicts and communal riots. It hurts both hosts and displaced communities (Cernea 2000: 3666–3667; Hemadri 1999: xx–xxi).

The development induced displacement sometimes become more complex, when they cross borders to eke out their livelihoods. It could have both positive and negative implications. Positive contribution could be the positive changes of economy of the host countries, while the negative consequences could be arms trafficking, drug smuggling, the trafficking of women and children over which many host governments have little control. In such situations, security concerns are magnified when host country fails to provide sufficient assistance to the large-scale migrants. Thus, host states perceive protracted refugee situations as posing direct and indirect threat to their national security and regime survival. However, not all refugees are seen as threats. In this sense, the importance of ethnic and religious affinity cannot be overstated. If host community perceives the incoming refugee as 'one of us' then positive and generous conception of distributive justice will apply. Conversely, if refugees are seen as members of an 'out group' they are likely to receive a hostile reception. In this case, there is a threat to the balanced multi-ethnic societies and also destabilize the political balance of power (Loescher and Milner 2005: 30–34).

The exiled communities or the uprooted populations are particularly vulnerable to statelessness, especially when their displacement is followed or accompanied by redrawing of territorial boundaries. Statelessness is not only a source of human insecurity and cause of forced displacement, but may also pose a threat to national and regional stability. Absence of citizenship is one of the important features of statelessness, generating tensions and even violence between different states and communities. The Universal Declaration of Human Rights unequivocally states that 'everyone has the right to nationality' and that 'no one shall be arbitrarily deprived of his nationality'. Thus, humanitarian organizations play a vital role in averting such situations (UNHCR 1997–98: 225).

The Kaptai dam constructed in 1963 in Bangladesh (earlier known as East Pakistan) submerged an area of about 400 square miles including 54,000 acres of cultivable land. About 90 miles of government roads and 10 square miles of Reserve Forest also went beneath the water. The lake took away 18,000 families and displaced 1,00,000 tribal people. The insecurity of livelihoods, accompanied with inadequate compensation and rehabilitation forced 40,000 displaced to migrate across the border of India and remained as the stateless citizen in the state of Arunachal Pradesh, Mizoram and Tripura (Sopher 1963: 347–348; Mohsin 1997: 102–103).

There are clear parallels between refugees and displaced persons. Both groups lack the protection of their governments. The root causes of both are similar and solutions for one are often interlinked with the other. Yet, the fact that one group crosses the border and another does not or cannot make a significant difference to their situation under the international law and the United Nations in regard to response to their plight. For the refugees, there is the international protective mechanism under the UNHCR mandate and a specific body of law to address their needs. The needs of the internally displaced remain to be addressed largely with the general provisions of human rights law and humanitarian law, measures and mechanism (Sen 1998: 182–183).

Internally displaced persons are entitled to enjoy full equality, the same rights and freedoms under domestic and international law as do the rest of the country's citizens. But in practice they rarely enjoy such rights and freedoms. Although the displaced are frequently forced to flee their homes for the same reasons as do refugees, they remain within national territory and come under the jurisdiction of their government. The government primarily bears the responsibility to protect them and giving them essential needs. However, the governments are unwilling or unable to guarantee the basic rights and meet the basic needs of their internally displaced citizens. Sometimes inter-governmental organizations, their specialized agencies and non-governmental organizations assume the role on an ad hoc basis (Deng 1998: 3–4).

It is well established that all individuals are endowed with basic human rights which are inherent attributes of human dignity and are recognized by the international law to protect them. States, in turn, are obliged to ensure respect for those universally recognized human rights which are essential to their survival, dignity and well-being of all persons subject to their jurisdiction. There are three sources of international legal standards: human rights law, which is applicable to all situations; humanitarian law, which is applicable in all situations of armed conflict; and refugee law generally applicable to the refugees but little chance of application to international displaced persons (ibid.: 3–5).

Resettlement and Rehabilitation

Before going into details about the resettlement and rehabilitation for the displaced people, it is very much essential to understand the clear meaning of terms like 'compensation', 'reparation', 'resettlement' and 'rehabilitation', since people often use these terms interchangeably and it is largely due to lack of academic precision. To be clear here 'compensation' understands as packages, cash or kind, either directly or indirectly, for the displaced. 'Reparation' refers to the acknowledged losses not only of assets but also of livelihoods and common resources, shelter and habitats. 'Resettlement' is understood as the packages and processes provided in new resettlement sites in addition to the compensations. Finally, 'rehabilitation' is seen as packages and processes provided in addition to those for compensation and resettlement in order to ensure the relocates better standard of living. Therefore, it is understood that rehabilitation is inclusive of compensation and resettlement (Hemadri 1999: xiii).

The main objective of the resettlement is based on collective negotiations with the affected people, making plans and implementing it as a development project to restore pre-project income and living standards. It is not only confined to physical relocation of the oustees but also aims to improve the quality of life by raising the living standards beyond the pre-project levels. The process of resettlement as a development project generally considered over a minimum period of two generations, which include not only protective measures, but also the provision of new rights, resources and strategies for the sustainable improvement of both subjective and objective indices (Bartolome et al. 2000: 27). There are two strategies (land based strategy and non-land based strategy) generally in practice for the socio-economic re-establishment for the oustees (ibid.: 31).

Land Based Strategy: It includes the adequate financial compensation for lost property and providing economic opportunities to re-establish their agricultural based livelihood sources. It is encouraged basically in Asia, Africa and Latin America, where most of the people are dependent on land for livelihood (ibid.: 31). The strategy of land for land is particularly more important options for the indigenous people and especially for women, for whom land is the only sustainable resource base. Land is the integral part for them and its separation can be traumatic. It almost provides the original habitat, and protection against land alienation. The land based strategy includes initiating and financing land reclamation activities, irrigation schemes, tree crops development, commercial or social forestry, vocational training, off-farm employment, and other kinds of lasting income generating activities (Cernea 1998: 184).

Compensation takes place in the form of cash or kind for the replacement of housing, land and other assets in the process of resettlement. In many cases, giving compensation to the displaced becomes more problematic. It often poses legal, social, political and economic problems and raises many questions. Most regulatory frameworks are inadequate to deal with compensation for customary rights or loss of opportunities for employment. In many cases, narrow definitions of property rights become obstacles to recognition of assets. Majority of the people affected by the development projects are from the poorest sections of the community and often have no legal title to the land or premises they occupy. The displaced generally the indigenous or tribal people and particularly the sharecroppers, agricultural labourers and workers lose their main sources of income (World Bank 1998: 30; Fernandes and Thukral 1989: 86–87).

In many projects, women are exempted from compensation of land. Most parts of Asia and Pacific are patriarchal society, dominated by male. The payment of compensation to those legal titles is intrinsically gender biased. Because land and property are mainly registered in the name of male and women are usually excluded from receiving compensation (Asian Development Bank 2003: 3). Even if some people have legal title of their land, they do not get fair compensation, which is much below the market value. Although, agricultural development is the primary goal of land for land resettlement, it has not been impressive in many parts of the world. This failure has been primarily due to lack of cleared land, lack of time, resources, marketing linkups, managerial capacity and heavy cost of land reclamation (Wet 1999: 9–11).

It has been estimated that, around 40 million people in India got displaced during the period 1950–2000 and among them majority are the tribal or indigenous people, who are landless, asset less, enjoy customary rights over land. Although under Article 21 of the Fundamental Rights, Constitution of India, it is clearly enumerated 'every person is entitled to a quality of life consistent with his human personality. The right to live with human dignity is the fundamental right of every Indian citizen'. But this is not being strictly followed in letter and spirit (Mahapatra 2000: 121–134). Legalizing customary and land tenure is most needed for successful land for land resettlement, which prevent alienation and consolidate rights. In Uganda, it has been legalized under the Land Act no. 16 of 1998, which has set a legal framework for gender equity in access to land rights over land. Article 237 too has empowered the Ugandan citizens for legal title over their customary land (Bartolome et al. 2000: 32).

Sometimes the resettled poor people do not get any benefit from the land based strategy of resettlement. In many cases, to protect the newly settled farmers, the government give agricultural loans at a nominal rate of interest, but corrupt practices by some officials in the loan granting agencies discouraged the displaced from taking government loans. Hence, government initiatives towards new occupations and rehabilitation have produced indignation and bitterness rather than mitigating of displaced grievances. The experience of land based resettlement in many reservoir projects is not very encouraging. It has been argued that unavailability and scarcity of land, poor project planning, lack of effort to identify land reserves and lack of political will of governments for providing lands are the major causes of unsuccessful resettlement (Cernea 1998: 184).

Non Land Based Strategy: Land based strategy is the important source of both livelihoods and food security for millions, while non-land based strategy is the diversification of the livelihood sources to other activities. The increasing pressure of land due to population growth and the increasing costs of agriculture, coupled with the rapid acceleration of economic activities compel other alternatives to grab new economic opportunities. It includes job creation for oustees through creation of industrial and service sectors and give them vocational training of different skills (Bartolome et al. 2000: 32). There are very few such success implementations of non-land based strategies. For example, an innovative and successful approach was taken at the Saguline Lake in Indonesia through the promotion of *floating net aqua-culture* as a productive activity for displaced families who lost their rice lands to the reservoir. It was estimated that about 1,500 families got benefit from this option within five years after reservoir impounding (Cernea 1998: 185). Similarly, the case of Xiaolangdi dam in China is another success example that can be mentioned here. The Government of China developed land and made the provision of 11,110 hectares of land for irrigation, relocated 252 small industries and mines and established 84 new industries, giving 25,500 new jobs to the displaced. This clearly underlines the importance of an integrated strategy of non-land based strategy to protect and ensure livelihoods (Jing 1999: 16). However, most resettlement programmes have not laid emphasis on the creation of non-land based economic avenues. Claims have been made that rehabilitating the displaced by giving jobs is quite an uphill task. Most of these jobs created by these projects are temporary in nature and many of the jobs need skilled personnel; government pays little attention or no effort to place them in the new project. Thus, exodus of skilled labourer is obvious in this area and the scheme of a job for an adult in a family is an improvement, but can't really serve the purpose in the long run. Getting a regular job in the project remains the single most important factor influencing the whole process of rehabilitation and it's largely due to lack of governments, commitments (Sharma 2003: 910–911).

Resettlement Experiences of Various Countries

The experiences of involuntary resettlement in many of the countries are mixed bag of success and failure largely due to the absence of rehabilitation policies of those concerned countries (Asian Development Bank 1995: 4). For example in India the absence of statutory rehabilitation law or even a national policy, there is no legal imperative for state governments or project authorities to integrate comprehensive rehabilitation planning into the planning of a project.

In India, it has been pointed out that it is the responsibility of the State Government to resettle the displaced and also the duty to make rehabilitation policies. In case of Ukai dam, the Government of

Gujarat, within five years, passed 20 different resolutions, recommended by various departments for the making of rehabilitation policy. During the 1960s, the Government of Andhra Pradesh had rehabilitation policy and it declared its willingness to rehabilitate the oustees made of Nagarjuna Sagar Project in the 1960. However, those promises eroded in 1965, and got away of an elaborate rehabilitation policy in case of Pochampad and Srisailam in 1965 and 1977, respectively. In India, different state governments have their different rehabilitation policies and keep changing according to the time and situations (Hemadri 1999: x). However, there are few experiences of improved resettlement in India, especially after the intervention of World Bank and its new resettlement guidelines. About 40,000 households and 2,40,000 people were successfully rehabilitated by the operation of World Bank in Karnataka, who were displaced by two dam projects of upper Krishna.

There are also mixed response of resettlement and rehabilitation in case of China. In case of Shuikou dam, the Government of China resettled 2,00,000 displaced people by deploying government employees aggressively, developed reservoir fisheries, fruits and timber trees, and township and village enterprises and even recruited the foreign investors to establish factories to employ oustees. It increased 44 per cent of their income within four yeas. The same experience has also been experienced in case of Yantan dam. Such successful experience were made largely due to these projects were funded by World Bank and resettlement and rehabilitation were made according to its guidelines (World Bank 2000: 1–4). On the other hand, there are also various evidences of failure cases in China. One among those is the Three Gorges Dam project (1994–2009), which is under construction on the Yangtze River in Central China, one of the largest dam developments in human history. It is uprooting 1.2 million people and it has become a difficult task for the government to have proper rehabilitation (Heming, Waley and Rees 2001: 195–204). It will affect 1,380 industrial enterprises and relocating those industrial enterprises has created numerous problems (Jing 1999: 21).

It can be stated that public agencies are typically limited in their capacity to handle resettlement and in most cases, they mishandled or ignore it. Sometimes the governments at the higher level are committed to good resettlement, but it is relatively less at the ground (resettlement officer) level. The increased differences between higher level and field level officials therefore frustrate the resettlement process. Many a times, discrepancies are made within the government owing to complex administrative structures. For example in Brazil, although the implementing agencies intended to provide the infrastructure and services needed to resettlers, often the federal government don't or could not allocate the funds to do the same in time (World Bank 2000: 1–2).

Question of Justice

Explaining the adverse implications of involuntary displacements, it is expedient to explain the John Rawl's 'general conceptions of justice' with regard to development projects – such as dam projects – and its impact of displacement is a utilitarian nightmare. It argues that, the project, which promotes greater good for greater number impoverish 'oustees' and compel them to sacrifice for 'greater good' is intolerable. The scheme or system, which creates advantage for some parts of community, is unjust, unless it creates benefits for all. Injustice, he advocates 'is simply inequalities that are not beneficial to all'. Rawls's general conception of justice enables us to recognize some of the problems encountered by 'oustees' as injustices. An unsuccessful resettlement scheme creates new inequalities by depriving them not only of 'income and wealth' but also of social goods in two other ways: 'liberty and opportunity' and 'social bases of respect' (Rawls 1971, Drydyk 1999: 1–8).

Income and Wealth: Displacement and faulty resettlement impoverish people by degrading the 'assets and resources', which they are formerly dependent on for their own subsistence and income. These include the five categories of 'impoverishments' (landlessness, joblessness, homelessness, marginalization, loss of access to common property resources) out of eight categories that Cernea has addressed. Losing these sorts of assets and resources impoverish people, what Rawls considers as 'social goods,' subject to the general conception of justice (ibid.).

Liberty and Opportunity: The development induced involuntary displacement curtails oustee's liberty. However, this involuntary displacement and the terms of resettlement and rehabilitation can be mitigated by consultation and negotiations by the state and the displaced communities and arguably the relocation would not be involuntary at all and no liberty would have been lost. It has been observed that many government agencies decide in secrecy on displacement needs, amounts, procedures, terms and deadlines, without the consultation and participation of those, whose vital interests are at stake. Rawlsian perspective advocates that liberty and opportunity are social goods subject to the general conception of justice. It is one of the social goods in which inequalities are not to be created, unless, they work to everyone's advantage. Therefore, it raises a standard for states and project authorities to consult with potentially displaced communities in order to find out the best possible solutions (ibid.).

Bases of Self-Respect: Rawls understanding of self-respect means one's life plans and his conception of good life, which develops his abilities and his conception of good, and his deeds is appreciated, esteemed and enjoyed by other people. He believes that people's belief of their own conception of good life is subject to the general conception of justice. From this perspective, social disruption caused by displacement is seen as unjust. It addresses two categories of 'impoverishments' (homelessness and social disarticulation) of Cernea. In the broader sense, the first one refers to distortion of group's cultural space and identity or cultural impoverishment. On the other, the second one refers to the dismantling of communities' social organization structure, dispersal of formal and informal networks, associations, local societies, etc. This undermines the degree of support, from where people derive their life plans and notion of good life. Depriving people from their self-respect and putting them in a new social scheme (resettlement and rehabilitation) is, therefore, unjust (ibid.).

Uneven Development

No doubt dams are made for the purpose of development and national prosperity and in this process many have to sacrifice for the sake of national interests, as it is earlier vividly described. However, observing closely, the benefits made by dams leads to uneven development and strengthen capitalism. Most of the time, it fulfils the interests of large farmers and industrialists.

Irrigation Lead to Concentration of Capital: Many rightly believe that irrigation made by dams helps in strengthening the capitalists structure, largely due to the extreme inequality in the distribution of land in the command area which gets further polarized with the introduction of irrigation by big dams. Although, land reforms were made in certain parts of the world, end of landlordism remains in half way. Many believe that the land reforms were not directed towards an egalitarian distribution of land, rather to smooth out the tax collection procedures.

It is obvious that the big landlords are the first to hear about an irrigation project and, therefore, many lobby for them with their political representatives. Once it is clear that the project would come through, they start to buy lands from the small and marginal farmers at a price much below the rate after the irrigation project comes through. Even before a project has been passed, the big landlords manage to increase their landholdings and eke out the small farmers.

In the next stage, they manage to divert the irrigation canals to their fields with political and other influences. They also see the activities of land leveling and shaping of land is done properly on their land to provide smooth irrigation. Such processes brought discrimination with regard to the intensity of irrigation in the command area. The secure water supplies are able to crop their land three times a year and, on the other hand, many rely on one time irrigation. Intensity of irrigation primarily depends on the timing of water deliveries. It makes meaningful, when farmers really want water at particular times. Deliveries outside these times may be of little value and may be even harmful. The farmers, therefore, adopt various tactics to get water when they want it. The connection with irrigation bureaucracy made them easy to get water, whenever they desire and, on the other hand, it reduces the costs of obtaining it.

The command area can be divided into three parts – Head Reach Area (HRA), Middle Reach Area (MRA) and Tail End Area

(TEA). It is usually the HRA that get good water flow supplemented by efficient land leveling. In the MRA, the land leveling is not properly done, due to shortage of funds and also because of carelessness of engineers as the project is nearly complete and there is hardly any inspection. They, however, receive water but not as well as the HRA. The TEA is the sufferer as the financial allocations have exhausted and there is hardly any water left due to siltation, water loss and intensive use of water in the HRA and MRA. Such common features made abundance availability of water in the HRA (especially the small minority of rich farmers) in time and they adopt water intensive cropping pattern which is commercially profitable irrigated crops. On the other hand, in the TEA, the mass of small farmers have to adopt a rain-fed cropping pattern, treating canal water as 'an intermittent blessings, when it come', but not to be relied upon. This is how the large dams lead to concentration of land and capital among the powerful. Similar conclusions have been drawn in case of Sreeram Sagar Dam in Andhra Pradesh. A micro-level study by Ali in 1986 (cited in Singh 1990) of the dam shows that the land holdings pattern is a skewed one, where the minority owns the majority of land. In this case, big farmers who were only few in numbers possessed more than 64 per cent of land. The average holding of land size is higher in HRA as compared to MRA and similarly they were higher in the MRA than in TEA. Study by Thorner in 1962 on Sarda Canal (cited in Singh 1990) reported that a small minority of rich farmers got as much as water as they wanted and were able to adopt profitable irrigated crops, while the mass of the small farmers had to adopt a rain-fed cropping pattern treating canal water as 'an intermittent blessing, to be welcomed when it comes' but not to be relied upon. The study of Velde in 1971 (cited in Singh 1990) shows that the distribution of water in part of the Bhakra canal gives evidence that landholdings closer to the canal outlet have a better standard of water service, and that big land holding receive more water per acre than small holdings. Big farmers' activities too directed towards the market and, thus, they shift to cash crops. In India, one-third of dams are in Maharashtra alone, where sugarcane farming has been thriving and it requires 10 times more water to irrigate sugarcane in comparison to kharif and seasonal crops. This shows how large dams lead to the concentration of capital among the powerful. To remove such discrepancies in distribution of water and proper land-levelling works, popular opinion has been voiced by farmers' organizations in large irrigation projects. It has also been argued that as to why irrigation water, provided by public funds could not be for all. It is claimed that there are 'very good opportunities for land distributions at the initial planning stage, before project completion'. However, these issues are put under the carpet (Singh 1990: 568–571).

Diversion of Water and Decreasing Area of Irrigation: In the recent years, it has seen that, a large quantity of water originally allotted for irrigation has been increasingly diverted to non-irrigation purposes. Rapid urbanization, industrialization along with changing lifestyle of the masses, puts enormous pressure on the government to augment water supply for urban and industrial use. Since the availability of water is limited, any reallocation will have to come from irrigation water, which is reducing the overall availability of irrigation water and resulting in reduction in area irrigated. The farmers at the tail-end area have been facing extra-marginal burden today in most of the command areas and, on the other hand, it has slowed down the expansion of canal irrigation, causing one of the responsible factors of food insecurity (Narayanamoorthy and Venkatachalam 2011).

The increasing diversion of water allotted for irrigation to non-irrigation purposes have been largely taking place in developing countries in the 1990s, triggered by the process of liberalization. The large amount of Foreign Direct Investments (FDIs) and the ongoing process of industrial activities in the peripheral areas to raise the economic growth of nations have been forcing governments to divert water, which is violating water 'user rights' of the farmers (ibid.).

Unavailable reliable water sources for non-irrigation purposes, states are trying to 'steal' the water from the dams and tanks for private purpose, which are originally constructed for the irrigation purposes. One can see the example of the State of Maharashtra in India. Some two decades back, Pune city was drawing about 4 tmc ft of water from Khadakwasala dam, but the drawl level reportedly has increased close to five times now. This increased drawl is already creating lots of hardship to the tail-end farmers of the Khadakwasala dam (ibid.).

Given the fast decline of unutilized irrigation water potential and increased competition for irrigation water from non-agricultural sectors, agitations and protests of farmers and their conflicts with government have been going on in many parts of the nation and it will continue in the future too. One such case can be mentioned here. The Government of Maharashtra was reportedly planning to divert more water from Pavna dam situated close to Pune city to Pimpri Chinchwad Municipal Corporation, which is one of the richest municipal corporations in India. The farmers protested against such move made by the government, who are historically relying on this water to maintain their livelihood. Interestingly, the movement and protests was suppressed by the mighty police forces of the state and three innocent farmers were killed by their bullets (ibid.).

Discussing about the 'user rights' over water, it differs from one country to another. In advanced countries like the US, the water transfer from agriculture to other sectors takes place with the consent of the farmers who are adequately compensated for sharing their water. The farmers are provided with 'use rights' which they can transfer to other sectors, if transfer benefits them adequately. The transfer takes place purely on voluntary basis. Government cannot transfer irrigation water without the permission of the farmers. But, in the Indian context, the property rights over water are still with the government. Farmers, despite property rights over land, could gain such right for water only if water reaches either the sub-canal or the field. The government can take away water from the main source whenever required. Moreover, farmers are also asked to pay for water even if they don't get water, since their land is demarcated under command area of the reservoir (ibid.).

Distribution of Electricity: It has been seen that most of the electricity is generated through hydropower or thermal power stations, which are located in the rural areas and which get their resources after displacing and encroaching upon the resources of local communities. But, it is very unfortunate that, among them, very few get the benefits. There is hardly allocation to the weaker section to set up cottage industries and any alternatives to promote their source of livelihoods. Instead, electricity is diverted to the national or regional grid which meets the requirement of the industrial houses, urban dwellers and lastly the rural areas. Government does not generally provide electricity to all the villagers and households; it has its centralized tendency and bias towards the economically powerful to fulfil their interest first. Large dams have enough potential to alleviate social inequalities that exist in the society, but actually it doesn't happen. On the contrary, those further aggravate the already skewed social structure in favour of the socially, economically and politically powerful (Singh 1990: 568–570).

Protests and Movements

Protest and participation, as distinct and contradictory forms of action, can be complementary in a democracy and protest movements may contribute to the legitimacy of democratic governance. Protest reflects the key aspect of political life – the relationship between the rulers and the ruled. Protest movements are the part of political participation, and they put pressure on the state authority to resolve the crisis faced by the ruled. Protests primarily originated due to the disagreement over any specific issue as opposition to the particular government policies. The type or the form of government in a country affects the number and intensity of protests. Protest and movements in democratic form of government are more extensive and less deadly than autocratic form of government. The structure and ethos of democratic regime better address and respond to the protests, armed rebellion or insurgency than the autocratic form of government (Swain 1997: 830–832).

Opposition to big dams has a long history since its beginning of construction in the ancient times. But in recent years this has come to command wider attention. As building of dams excelled, especially after the 1950s, opposition to dams has become more widespread. Protests to dams were made in the northern countries, especially in the United States, led successes for campaigns against large dams. They stopped the 175 metre-high Echo Park on a tributary on the Colorado River in the 1950s and two dams planned for the main stem of the Colorado River in the Grand Canyon in the following year (World Commission on Dams 2000: 18–19). Similarly, in the developing or Third World countries, protests and movements were staged against big dams. There were stiff resistance to Pong dam, Koel Karo, Tehri, Sardar Sarovar dams projects, etc. in India (Centre for Science and Environment 1999: 133–137), Kariba dam in Pakistan, Arun III in Nepal, Three Georges dams in China, Pak Mun in Thailand, Ilisu in Turkey, Ralco in Chile, Epupa in Namibia, etc. (World Commission on Dams 2000: 18–20).

Until the 1970s, the opposition to dam was confined to concerns of government agencies, engineers and economists. But in the late 1980s, environmentalists and sociologists played a greater role influencing the planning process and by the mid-1990s, the involvement of affected people and NGOs in the process became more significant. The last few decades have also seen a number of significant policy steps in response to civil society demands and changing values. More recently, the trans-national coalition of ideologies and advocacies strengthened the oppositions and it put major impact on the planning and policy-making on individual dams. Such important milestones include the World Bank's withdrawal from the Sardar Sarovar Project in India in 1993 (Centre for Science and Environment 1999: 137) and Arun III in Nepal in 1995 (World Commission on Dams 2000: 19).

Note

1. In India, the power of 'eminent domain' was recognized on the principle that the sovereign state always can acquire property of a citizen for public purpose without the owner's consent, which is made by the framers of the Indian Constitution to the people of India. The power of eminent domain conflicts most with the Constitutional imperatives contained in the part XVI of the Constitution of India, designed to protect Scheduled Tribes. Most state legislatures have passed elaborate statues to protect tribal land owners from alienation of their lands, but paradoxically no protection is extended to tribals for loss of lands.

3

Dams and Development in India

Development and Planning: Rationales of Introduction of MPRVD

At the time of Independence, India was stranglehold of stagnating per capita national income, static agriculture production, poorly developed industries, inadequate infrastructure, mass poverty and extreme unemployment, etc. Among all these problems, food scarcity was worrisome to the whole nation. Indian agriculture was in deplorable condition. Its productivity had been declining and was generally among the lowest in the world. Despite the fact that nearly 70 per cent of the working population was engaged in cultivation; the country was not self-sufficient in food grains, and had come to depend on imports of food grains (Datt and Sundaram 2003: 510).

Additionally, the Partition of the country in 1947 worsened the agricultural situation further. In fact, the statistics indicate that Partition had placed Pakistan in a relatively superior position with respect to agriculture. Undivided India used to irrigate 70 millions acres of land, the largest irrigated area in any country of the world and three times that of irrigated land in the United States then. Out of this, India got 48 million acres or 68 per cent and Pakistan 22 million acres or 32 per cent. The proportion of irrigated area to net shown area is larger in Pakistan than that in India (Vakil 1950: 126–185).

In this situation, there was a need to put in huge and organized effort on a national scale to achieve substantial progress on the socio-economic front. Thus, planning was accepted as the key strategy of India's developmental efforts. The era of planned development was ushered in with the launch of the first 'Five Year Plan' in April 1951. It addressed the problems arising from massive influx of refugees, acute food shortages and mounting inflation. The highest priority was given to overcoming food crisis by raising food grains output, curbing inflation and the development of infrastructure. An important part of India's agricultural and industrial development is the utilization of the country's great water resources, particularly through multi-purpose river valley projects (Huyck 1953: 88–99; Rao 1952: 3–23; Swain 1997: 823–825).

The Central Water-ways, Irrigation and Navigation Commission, a government agency was set up to plan and promote such projects. It declared that only about 5 per cent of India's water resources were utilized – irrigating some 46.89 million acres or 19 per cent of the cultivated land, and producing about half a million of an estimated potential of 40 million kilowatts of hydropower. To harness these resources, there were 46 river valley projects being planned by both the central and state governments at a cost estimated at over four billion rupees ($300 million), intended to provide irrigation for about 8.7 million acres and production of about 1.1 million kilowatts of hydropower. This was target set for 1965 by the Planning Commission in India's first Five Year Plan. Some of the important river valley projects are Hirakud dam project in Orissa, Bhakra and Nangal project in Punjab, Tungabhadra River project in Tamil Nadu and Andhra Pradesh and Kosi River project in Bihar (Morgan 1952: 82–83). Like the First Five Year Plan, construction of dams was given high priority in India's each Five Year Plans and over 15 per cent of the total plan outlay was invested on dams. Till now, irrigation and hydropower generation have been the major objectives of water resource development in India (Datt and Sundaram 2003: 536).

Following the national development strategy and economic growth as its indicator, large scale industrialization and massive infrastructural development projects were set up to set India on the path of modernization and development. Immediately after Independence, a series of large dams were planned and built on the major rivers of India. The first Prime Minister, Jawaharlal

Nehru called dams as 'secular temples of modern India'. It promised to solve the problems of floods, hunger and starvation by providing irrigation and enhancing food production and providing much needed electricity for industrial development and following this strategy, now India became one of the largest dam building nations in the world. There are 4,291 dams in India and out of those 3,596 have been built and 695 are under construction (Patwardhan 1999: 1–12).

However, adoption of MPRVD in India was not solely triggered by it internal compulsions, rather, it was influenced by developed countries too, especially the US. The global model for MPRVD was pioneered in the United States in the early 1930s with the establishment of TVA after which a series of dams were constructed on the River Tennessee and its tributaries. The TVA experience with comprehensive river regulation encouraged other regions of the world to replicate the MPRVD in different river basins. The TVA model became the icon of modernity. The era was marked by the rapid decline of the European colonial powers and the emergence of the United States of America as the new hub of capitalist dynamism. Many governments adopted MPRVD immediately during their post-colonial phase with the help of American experts for its construction to speed up their economic development. In other words, Rohan D' Souza says that the MPRVD scheme was inextricably linked to the political economy of decolonization and nationalism (D'Souza 2003: 82–83).

Dams and Development

Flood Control: Flood plain areas are ecologically important wetlands and have provided competitive advantages for human settlement in the past. However, these plains are also prone to flooding. This leads to what is called 'flood hazard'. Defined simplistically, flood hazard is the threat to life and property, health, livelihoods and other valued resources presented by a body of water, which might rise and flow over land that is not normally submerged. Flood disasters are among the world's most frequent and damaging types of disaster. During the later half of the twentieth century, floods were the most common type of geophysical disaster, accounting for over 30 per cent of all disasters between 1945 and 1986. In terms of loss of life, they have proved to be the third most fatal of all forms of geophysical disasters globally. In the period given above, only earthquakes and tropical cyclones killed more people than floods. However, in the next decade (1986–1995), floods appear to have caused more deaths than any other geophysical disaster. During this period, 55 per cent (367, 000 people) of the deaths were caused by the floods alone. The estimation of deaths caused by flooding in 1999 was 55,360 and among them 20,000 had died in Venezuela alone, caused by mud floods and landslides. In terms of regional significance, flood related economic losses in Asia exceeded those in North America and Europe between years 1987 and 1996. Thus, flood hazard management is most essential for protecting people from such calamities (Green, Parker and Tunstall 2000: 2–4).

Managing flood hazards, construction of dams on turbulent rivers was widely accepted mechanism of flood management. The big dam projects increasingly became more popular and contributed to the formation of various informal international big dam regimes. The wave of decolonization and state formation contributed to the spread of these institutions to other parts of the world and especially in the Third World. The spree modernization and the attendant strategies of rapid economic growth adopted soon after the Second World War accelerated the global dam construction rate. India now ranks among the most important dam building nations in the world. It has so far constructed 4,000 dams and 40 per cent of dams under construction in the world today are reportedly being built in India. About 13 per cent of all the large dams in the world, in more than 75 countries, have flood management functions (Khagram 2004: 5–7; World Commission on Dams 2000: 8–10).

Irrigation: Agriculture plays a greater role in the economic and social development of all developing countries. Even though it is the developed countries with industrial base that still account for a significant part of the world agricultural production. In this context, irrigation becomes crucial for food security as it is vital for agriculture in many regions of the world particularly in the tropics, where rainfall is seasonal and unevenly distributed. Irrigation is seen as a vehicle for development, poverty alleviation, rural development and achieving food insecurity. Today, about one-fifth of the world's agricultural land is irrigated and it accounts for about 40 per cent of the world's agricultural production. Many countries, including China, India, Egypt, Indonesia and Pakistan, rely on irrigated land for more than half of their domestic production. It was during the first years of green revolution in the 1960s that the area under irrigation expanded dramatically in many of these countries. This led to significant yields and brought down food prices. As a result, today four countries – China, India, United States and Pakistan – account for more than 50 per cent of the world's total irrigated area (Sanmuganathan 2000: 22–57; Gritzinger 1987: 14–15).

India is a predominantly agricultural country, and agricultural performance is fundamental to its economic and social development. Thus, development and management of water plays a greater role in socio-economic well-being in India. People of various regions of the country get both direct and indirect benefits from the large scale investment in water infrastructure like multi-purpose dams, canals, small check dams, tube wells and pump sets (Bhatia 2007: 101–102).

In the early years of Independence, Indian agriculture was in deplorable condition. Except in certain areas, it was dependent on rainfall and the vagaries of monsoon. Agricultural productivity in India had been declining and was generally among the lowest in the world. Despite the fact that nearly 70 per cent of the working population was engaged in cultivation; the country was not self-sufficient in foodgrains, and had come to depend on imports of food grain (Datt 2003: 510).

Additionally, the Partition of the country in 1947 worsened the agricultural situation further, as mentioned earlier (Vakil 1950: 126–185). Therefore, the Indian planners, right from the beginning, realized the need to attain self-sufficiency in foodgrains as one of the important goals of planning. After Independence, in the era of planned development, particular emphasis was laid on water development for irrigation to bring food security (Mohile 2007: 10). In all the Five Year Plans, expansion of irrigation has Formed an important component of each of the plans. Soon after Independence, multi-purpose river valley projects and other dams and canals were considered essential for meeting India's critical requirements of irrigation for agriculture, electricity for industries and controlling floods in turbulent rivers. Dam construction was given high priority in India's economic plans and over 15 per cent of the total plan outlay was invested on dams and canals. Till now, irrigation and hydropower generation have been the major objectives of water resource development in India (Datt and Sundaram 2003: 536).

In the six decades of Independence, the Government has invested large amount of resources on major, medium and minor irrigation projects. Since the First Five Year Plan, considerable importance has been attached to the provision of canal irrigation. In addition, the Government has also spent on other sources of irrigation like wells, tube wells and tanks, etc. Over time, canal irrigated area increased from 8.3 m.ha in 1950–51 to 17.1 m.ha in 1996–97 (Datt and Sundaram 2003: 332). Although the irrigation development has not been impressive in this period (less than 0.8 m.ha per annum), the Net Irrigated Area (NIA) has increased from 20.9 m.ha in 1951 to 57.2 m.ha in 2000 (Sekhar 2007: 47). This has greatly contributed to the increase in foodgrain production from 51 million tonnes in 1950–51 to 203 million tonnes in 2001–2003 (Datt and Sundaram 2003: 332). Agriculture contributes to 26 per cent of GDP and 60 per cent of employment in India. Besides, it is the primary source of livelihood in rural areas which account for 72 per cent of India's population. The performance of irrigated agriculture contributes more than 55 per cent of agricultural output and it is one of the most influential factors on the objectives of growth, employment generation, food security and poverty reduction (Bhatia 2007: 101–102). It also accounts for 14.7 per cent of the total export earnings and provides raw material to a large number of industries like textiles, silk, sugar, rice, flour mills and milk products. This shows that widening irrigation has large potential to improve crop yields and bring about a marked increase in farmer's income, poverty reduction and food security (Sekhar 2007: 47).

Hydropower: As per assessment made by the Central Electricity Authority (CEA), India is endowed with economically exploitable hydropower potential to the tune of 1, 48, 700 MW of installed capacity. The basin wise potential is: Indus basin (33,832 MW), Ganga basin (20,711 MW), Central Indian River System (4,152 MW), Western Flowing Rivers to southern India (6,430 MW), Eastern flowing Rivers to southern India (14,511 MW), and Brahmaputra basin (66,065 MW). Now, the electricity sector in India had an installed capacity of 202.98 Gigawatt (GW) as of May 2012, the world's fifth largest (India Hydro Energy).

Thermal power plants constitute 66 per cent of the installed capacity, hydroelectric about 19 per cent and rest being a combination of wind, biomass, waste-to-electricity, and nuclear. India is blessed with immense amount of hydroelectric potential and ranks 5th in terms of exploitable hydro-potential on global scenario. The public sector has a predominant share of 97 per cent in this sector. National Hydroelectric Power Corporation (NHPC), Northeast Electric Power Company (NEEPCO), Satluj Jal Vidyut Nigam Limited (SJVNL), Tehri Hydro Development Corporation and NTPC-Hydro are a few public sector companies engaged in development of hydroelectric power in India. Besides those, Bhakra Beas Management Board (BBMB), an illustrative state-owned enterprise in north India, has an installed capacity of 2.9 GW (Wikipedia: Electricity Sector in India).

Power generation in India has come a long way from about 1,000 MW at the time of Independence (August 1947) to about 1,60,000 MW as on 31 March 2010 (end of Financial Year). The share of hydropower in this growth, in these over six decades, has also been impressive as it increased from about 500 MW at the time of Independence to about 37,000 MW as of March 2010. The present level of hydropower exploitation is about 25 per cent of the ultimate installed capacity estimated at 1,50,000 MW (ASME Digital Collection).

Water Supply to Urban Cities and Industrial Needs: Rapid growth of urban population has been the character of Indian urbanization. The urban population in the country has been increasing manifold since Independence. However, the decadal growth rate in urban population in 1981–91 has decreased to

36.4 per cent, as compared to 46.1 per cent in 1971–81. The reasons for rapid urban growth rate have been largely attributed to natural increase in rural-urban migration, especially since 1961. The rapid urbanization and growing population in many of the cities in India has demanded increasing safe drinking water. In case of India, the water sources in urban areas have too depleted, polluted or destroyed their local sources of water. Lack of provision of adequate minimum water for vast proportions of poorer segments on the one hand, and wasteful use without paying even cost prices by more prosperous segments on the other hand is typical picture of most urban areas. Thus, having exhausted, mismanaged, polluted the local sources of water, and after continuing to neglect the local water potential, the urban areas increasingly look at existing large dams or new proposals of large dams to satisfy its water demands (South Asian Network on Dams, Rivers and People 1999).

Though water supply forms a minor component of the 3,303 large dams constructed in India since Independence, in recent years it assumes significance with rising demands of water. Various dam-based water schemes have been made for cities and metros in India to quench their water demands. Few examples can be cited here: New Delhi on Bhakra storage, Tehri dam, Renuka dam and Kishau dam; Hyderabad on Nagarjunasagar, Osmansagar, Himayatsagar and river Krishna; Ahmedabad on Sardar Sarovar dam and Dharoi dam; Chennai on Telugu Ganga; Mumbai, Vitarna scheme, Bhatsai project and Usgaon dam; Bangalore on K.R. Sagar, etc (ibid.).

Various dams in India too supply water for industrial needs. Damodar River Valley, the multi-purpose river valley development plan, comprising Maithon dam, Panchat dam, Tilaiya dam and Konar dam, has been providing 680 cusec of water to meet industrial, municipal and domestic requirements in West Bengal and Jharkhand (Wikipedia: Damodar Valley Corporation). The Upper Wardha dam, in the state of Maharashtra has been providing 2,50,00,000 litres of water per day to the water treatment plant for industrial purpose, which has been built near Amravati town (Wikipedia: Upper Wardha Dam).

Extent of Displacement and Rehabilitation

Indeed, contribution of dams in India is enormous, putting the nation in the path of development; but, the social and ecological costs of large dams were grossly underestimated and ignored. Large scale displacement has been justified by invoking the utilitarian logic that 'few people have to sacrifice for the national interest or goods'. However, continuous people's movement and work of researchers have brought the social, ecological and political considerations of large dams to the forefront in recent periods. One of the facts about dam building in India is the absence of reliable data based on performances as well as the impacts of large dams. The official data particularly are dismal due to lack of its authenticity and lack of concerns of the state towards displaced people and it reflects the dominant attitude of neglect. Thus, it prompted researchers to rely on independent estimates and it also varies from person to person. However, there is an agreement that dams are single largest cause of displacement accounting for about 75 to 80 per cent (40 million) of the total displacement in India (Patwardhan 1999: 1–12).

It has been pointed out that the majority of the displaced belong to the poor and deprived classes. Almost 40 per cent of the displaced are the Scheduled Tribes (STs) and 20 per cent belong to the Scheduled Castes (SCs). It is unfortunate that the tribal accounts for just 8 per cent of total population, but miserably shares 40 per cent of total displaced in India. Although the Constitution of India has empowered equal opportunity to all her citizens by providing special protection to the weaker section (SCs and STs), they are being marginalized in the process of 'national development'. Uprooted from their ancestral land, they are forced to migrate to urban areas in search of employment and become landless labourers. The deprived classes have been paying the prices for 'development' of the urban areas and large farmers by providing irrigation and electricity. Dams have been served as another instrument of dominant classes for appropriating the two most important natural resources–water and land from less powerful communities like adivasis in an unequal society like India (ibid.).

In fact, involuntary displacement of people is not new to India. It has been continuing since the British colonial rule and intensified in the decades of planned development. It is estimated that 2,13,00,000 people were displaced in the period 1950–1991. As per the government records, 75 per cent of those displaced are still not resettled and rehabilitated. This is, indeed, a sad reflection of the government's sincerity and sense of responsibility. Such involuntary displacement is traumatic. The experience of the last 50 years has demonstrated that both the project authorities and the administration as a whole have proved to be incompetent in reducing the trauma of displacement. The administration has violated the individual and group rights while implementing the resettlement plans. The degree of suffering experienced by those victims can not be compensated in terms of monetary values. Even if, in many cases, compensations are made for their lost properties, it is too late for them and they feel defeated and helpless. Almost all the developmental projects are set up in the remote peripheral area, where the tribal, ethnic minorities, pastoralist inhibit and, therefore, restoration of their socio-economic status has been rarely been successful. It's a cruel joke that after 60 years of Independence, the Government of India did not wish to promulgate a National Policy on Resettlement and Rehabilitation for the displaced people. It is the lack of awareness on the part of the country's policy makers and their unwillingness to rehabilitate the displaced (Fernandes and Paranjpya 1997: 1–34).

Environmental Costs: Construction of dams and its proliferation raised several important environmental questions in India too. Vast areas of lands get submerged under the reservoirs of dams. In India, it has been calculated that big river valley projects have swallowed 0.5 m ha of forests land between 1951 and 1976 – roughly one-tenth of the area which has benefited from irrigation. One of India's most controversial dam projects, Sardar Sarovar dam, its environmental impacts and net costs and benefits are widely debated. The World Bank was initially the funding agency of the project, at last withdrew in 1994. The Narmada dam has been the centre of controversy and protests in India since the late 1980s (Singh 1990: 568–571).

The Institutional and Political Context of Displacement

Land Acquisition: Law and Policy: The most important factor that displaces people in India is the prevailing Land Acquisition Act (LAA) of 1894. It empowers the state to enjoy unlimited power over land within its territory. It follows that the state has the right to invoke this right for 'public good' and the consequent compulsory acquisition of the land can not legally be challenged or resisted by a person or community. The doctrine of 'eminent domain' in India enjoys the power for the acquisition of land (Baxi 1989: 164–171; Hemadri 1999: xxxii–xxxiv; Fernandes 1999; Bartolome et al. 2000: 9; Ramanathan 1995; Sharma 2003: 907–910). It was introduced by the British government and it has been continuing in independent India too. It justifies the concept of 'public domain', by which the British Government was able to legitimize controlling over land. The main objective of the act was to acquire land for railways, expansion of trade routes, conversion of the forests and pasture land into plantations of tea, coffee, rubber, indigo, establishment of army cantonments and ordinance factories, construction of dams and canals, etc. and in these cases, the worst victims were the displaced, mostly the forest dwelling communities, tribal and the farmers (Fernandes and Paranjpya 1997: 8–9).

The Indian Forest Act of 1865, introduced by the British, too facilitated the acquisition of large forest areas for building railway lines. It established absolute control over the forests land in the name of public goods and converted them into state property. Thus, the community dependent on forests for their livelihood lost their source of earning and customary rights (ibid.). As discussed earlier about the power of eminent domain, which is meant to acquire private land for other public purposes is provided through article 31A and 298 of the Indian Constitution. Forests were initially placed on the State List of the Constitution of India, but the 42nd Amendment Act of the Constitution transferred the forests resource from the State List to Concurrent List, which empowered both the central and state governments and vested ultimate authority to make laws relating to forests. The ongoing process of centralization of administration by the governments and their control over forests land has further extended the process of displacement (Upadhyay and Upadhyay 2002: 12, 19).

International Finance and Dam Building: Construction of large dams in India is generally made by both central and state governments and it is supported by both domestic and international financial institutions. Among them, the most resourceful financial institution, the World Bank, started lending for large dams in India in the 1970s and gradually it became the largest foreign donor to public sector irrigation projects in India. Although, it's funding constitutes only 26 per cent, displacements made by those projects share 62 per cent of the total number of people displaced in India. However, the World Bank withdraws from financing dams in India in 1993, facing lots of challenges from the massive public campaigns against dams and especially from Sardar Sarovar, Tehri, Maheshwar, Omkareshwar, Tipaimukh projects, etc. in the 1980s and 1990s. The World Bank distanced itself from lending for large dams in India for over 10 years and came back again in December 2003, deciding to double its infrastructure lending for India. Getting rays of support from the World Bank, in 2003, the Government of India proposed over 160 new hydroelectric projects to double the current electricity generation in the country and most of these projects will be carried out by India's National Hydropower Corporation (NHPC). Besides World Bank, financial institutions from Sweden, Norway, the United Kingdom, Canada, Japan and Germany had provided loans to NHPC. Skandinaviska Enskilda Banken, Credit Commercial de France, HSBC, the Nordic Investment Bank, the Export Development Corporation, the Japan Bank for International Cooperation, ABN Amro, ANZ, Barclays, Natwest, Standard Chartered, Sumitomo and a syndicate headed by Deutsche Bank also extended loans for building hydropower dams in late 2003 (International River Network 2005).

Privatization and Market in India: In the 1990s, with the advent of New Economic Policy in India, it attracted large scale private investments and inflow of foreign investment demanding more lands in order to compete in the world market. India's moves towards privatization, inviting multinationals has further intensified the extent of displacement. Particularly the tribal communities are the worst victims in era of the new Economic Reforms. Greater purchasing power of the multinational companies makes the process of land acquisition and rejection of the draft National Policy for Resettlement, which recognizes some basic rights of the displaced people, much easier. This process is treating the natural resources as mere commodities, rather than recognizing the importance as means of livelihood for several communities. The state instead of safeguarding the rights of the citizens, has been acting as an ally and promoter of private companies without responding to the basic human rights. Sometimes, the state resorts to repression and brutal police action, when people demand for their rights (Patwardhan 1999: 12).

It is clearly visible that most of the states are acquiring land for private companies. For example, Orissa had acquired 40,000 ha of land for industries during the period 1951–1995, but plans to acquire 1,00,000 ha of land in a decade. Similarly, Goa had acquired 3.5 per cent of state landmass during the period 1965–1995 and plans to acquire 7.2 per cent of it in a decade. Andhra Pradesh has already acquired in five years half as much land for industry as did in 45 years. Similar quantities are being acquired in Jharkhand for the purpose of mining for foreign private companies (Fernandes 2004: 1192).

Resettlement and Rehabilitation Policy in India

There is ample historical evidence of involuntary displacement in the first half of the nineteenth century. A major wave of development took place between 1830 and 1842, largely due to colonial industrial policy of the British government, aimed at crushing the Indian manufacturer and handicrafts. It was a great setback to the indigenous textile industry, replaced by mass manufactured British textile in India. This phenomenon is vividly encapsulated during the period of Lord William Bentinck. The displaced population moved from one worksite to another as low paid daily wage labourer in the construction work. Even, most of them were forced to migrate as labourer to other British colonies, when they didn't get opportunity to work anywhere else. Later the involuntary development induced displacement was justified by introducing the Land Acquisition Act of 1894 (LAA) and it continued too in the independent India. However, there was no such provision made by the British Government to rehabilitate those displaced (Fernandes and Paranjpya 1997: 1–34).

Even no specific policy was ever formulated either by the central or the state governments to address the issue of involuntary displaced people, when the Indian state initiated development projects in different parts of the country soon after Independence (Pandey 1998: 9). After displacing about 40 million persons during 45 years of planned development, the Government of India has finally realized the need to formulate a rehabilitation policy for the displaced persons (Fernandes 1997: 35), when it founded that over 40 per cent of the displaced people during this period were tribals. Finally, for the first time in the post-Independence India, a rehabilitation policy was drafted in 1993. It was drafted under the pressure of both internal as well as external factors. The internal factors include: (1) the rise of protests and movements marked by a growing militancy, (2) national awareness of the problems of displacement, (3) and the demand for constitutional rights. On the other hand, the external pressures include: (1) environmental concerns raised by the Rio Conference, (2) concerns raised by Human Rights Conventions, (3) and the pressure comes from funders like World Bank (Fernandes 1997: 36; Fernandes 1998: 2703; Fernandes 2004: 1191; Ministry of Rural Development 2004; Kujur 2005: 138–151; Asian Indigenous Tribal People Network 2006a).

Protests, Movements and Growing Militancy: The importance of land as a livelihood resource cannot be overstated and the central slogan of the Naxalite movement has been 'land to the tiller', putting the poor in possession of land, and it defined much of their activity. Today in India, 58 per cent of the people engaged in agriculture, but they only contribute 18 per cent of the GDP. Forty per cent of rural households have no land or less than half an acre of land. It has been estimated that the number of landless rural families in the country is 1.30 to 1.80 crores and the number of small and marginal operational holdings has been increasing steadily over the years. In such situation, land acquisition by the government, imposing utilitarian logic does not get any support; rather, it accelerated protests and movements (Government of India 2008: 45–48).

Growing militancy, especially the Naxalite movement comes from the most deprived *dalits* (as known as Schedule Castes) and *adivasis* (as known as Schedule Tribes), who comprise about one-fourth of India's population. The former constitute 16 per cent and the latter 8 per cent and most of them (80% of *dalits* and 92% of *adivasis*) live in rural areas. The high levels of rural poverty not only put high distress among them, but also they suffer from multi-faceted oppression and denial of justice, social, legal and political rights (ibid.: 3–7).

Unrest and discontent are not new to tribal areas, nor is it just a post-Independence phenomenon. Over the last century, their political, social and economic life changed under the impact of the colonial administrative system, and the most significant of these changes has been the loss of command over their resources. In general, the contradiction between the tribal community and the State itself has become sharper, translating itself into open conflict in many areas. Almost all over the tribal areas, including Nagaland, Manipur, Tripura, Assam, Jharkhand, Orissa, Chhattisgarh, Maharashtra, Andhra Pradesh and Kerala, tribal people seem to feel a deep sense of exclusion and alienation, which has been manifesting itself in different forms and it thereby contributed to their disempowerment and deprivation. Apart from poverty, numerous forms of deprivations invoked them to raise protests and movements against Government policies. The most important among them are absence of self-governance, forest policy, excise policy, land related issues, multi-faceted forms of exploitation, cultural humiliation and political marginalization. Land alienation, forced evictions from land, and displacement made the important basis of protests and movements across the nation (ibid.: 8–10).

Much of the unrest and growing militancy of Naxalites is linked to lack of access to basic resources to sustain livelihood. The conflict pertaining to forest dwellers' rights to land and forest produce is a major source of unrest in large parts of the country. The Forest Conservation Act, 1980, made this position irreversible by declaring that no forest land shall be diverted to non-forest use without the permission of the Union government. The punitive provisions of the Act meant that eviction of *adivasi* occupants of forest land took place on a regular basis, resulting in considerable deprivation and suffering. Forest conservation has found a strange companion in industrial forestry. While the rights of the forest dwellers are severely restricted in the name of forest conservation, the forests are increasingly shaped to suit the needs of industry. This has resulted in the forest dwellers simultaneously losing access to land and to a variety of forest produce (ibid.: 11–15).

There is official database of persons displaced/affected by development projects However, some unofficial studies, particularly by Dr. Walter Fernandes, peg this figure at around 60 million for the period from 1947 to 2004, involving 25 million ha. which includes 7 million ha of forest and 6 million ha of other Common Property Resources (CPR). Among them, 40 per cent of the displaced/affected are tribals, 20 per cent *dalits* and 20 per cent OBC too. The resettlement record is also very dismal. Only one-third of the displaced persons of planned development have been resettled. As tribal areas are also rich in mineral resources, the mining projects proposed such as in Orissa, Jharkhand and Chhattisgarh threaten the very existence of tribal people. Protests and movements become an inevitable consequence of displacement, such as protests in Kalinganagar and Kashipur against Utkal Alumina in Rayagada district of Orissa (ibid.: 15–16).

Environmental degradation and social injustice are two sides of the same coin. Geographical distribution of fossil fuels, and metallic and non-metallic mineral reserves, shows a high degree of concentration in central and eastern India. The drastic changes in the existing land use from agriculture and forestry to various developmental activities, especially dam projects, mining and industrial activities has caused growing violence in the periphery. Environmental degradation resulting from such activities has created serious health problems for the local population. A large number of occupational diseases are caused by dust from mines and polluted air from industrial units. Uranium mining and processing near the Subarnarekha River has even caused radioactive pollution. The misery of the people get compounded from the consequential effects of illegal mining. The activities for exploitation of water resources directly create conflict with the local population. Large dams by changing the course of nature cause severe damage to the natural environment and rich biodiversity as they are located in ecologically sensitive regions (ibid.: 17–18).

Growing National Awareness and Demand of Constitutional Rights: According to the census of 2001, there are 84.3 million tribals, residing mostly in the forest areas in all the States of India except Punjab, Haryana, Delhi and the Union Territories of Pondicherry and Chandigarh. The architects of the Constitution, being conscious of the distinct identity of the tribal communities and their habitat, provided certain articles exclusively devoted to the cause of the tribal people, including Articles 244, 244A, 275(1), 342, 338(A) and 339. Following these provisions in the Constitution aimed at ensuring social, economic and political equity, several specific legislations have been enacted by the central and state Governments for the welfare and protection of tribal people and their tribal domain (ibid.: 8–9).

In the 1970s, a serious attempt to focus on the tribal population in the planning process was made in the form of a Tribal Sub Plan strategy. The process of bringing all tribal majority areas under the Fifth Schedule of the Constitution was also taken up. The 73rd and the 74th Amendments to the Constitution of India, followed by the Provisions of Panchayats (Extension to Scheduled Areas) Act 1996 (popularly known as PESA), brought in a new model for self-government in the Fifth Schedule areas of the country. Despite the plethora of development plans, programmes and activities initiated in the tribal areas, the majority of Scheduled Tribes still live in conditions of serious deprivation and poverty. The tribal people have remained backward in all aspects of human development including education, health, nutrition, etc. Apart from socio-economic deprivation, there has been a steady erosion of traditional tribal rights and their command over resources caused by large scale displacement (Government of India 2008: 9–10). Such violation of the basic rights of the affected people has resulted in growing interaction between local communities and activists from beyond the affected region, and the articulation of the rights and the injuries has been moulded in the process of this interaction. In the recent years, the right to resources is vigorously contested terrain in the human rights movements (Patel 2010: 469).

The Indigenous and Tribal People Convention 1989: The International Labour Organisation (ILO) Indigenous and Tribal Peoples Convention 1989 (No.169) is the major binding international convention concerning indigenous peoples and a forerunner of the Declaration on the Rights of Indigenous Peoples. Now it has been ratified by 20 countries. The first general, fundamental principle of Convention No. 169 is non-discrimination. Article 3 of the Convention states that indigenous peoples have the right to enjoy the full measure of human rights and fundamental freedoms without hindrance or discrimination, be it male or female. Article 4 guarantees enjoyment of the general rights of citizenship without discrimination. On the other hand, Article 20 provides for prevention of discrimination against indigenous workers too. In response to the various vulnerable situations of indigenous and tribal peoples, further Article 4 calls for special measures to be adopted to safeguard the persons, institutions, property, labour, cultures and environment of these peoples. In addition, the Convention stipulates that these special measures should not go against the free wishes of indigenous peoples. Indigenous and tribal peoples' cultures and identities form an integral part of their lives. Their ways of life, customs and traditions, institutions, customary laws, forms of land use and forms of social organization are usually different from those of the dominant population. The Convention recognizes these differences and aims to ensure that they are protected and taken into account when any measures are being undertaken that are likely to have an impact on these peoples. Article 7 states that indigenous and tribal peoples have the right to 'decide their own priorities for the process of development as it affects their lives, beliefs, institutions and spiritual well-being and the lands they occupy or otherwise use, and to exercise control over their economic, social and cultural development (International Labour Organisation).

Since its adoption, it has gained recognition well beyond the number of actual ratifications. Its provisions have influenced numerous policy documents, debates and legal decisions at the regional and international levels, as well as national legislation and policies. The Convention stipulates that governments shall have the responsibility for developing co-ordinated and systematic action to protect the rights of indigenous and tribal peoples. With its focus on consultation and participation, it has become a tool to stimulate dialogue between governments and indigenous and tribal peoples and has been used as a tool for development processes, as well as conflict prevention and resolution (ibid.).

Environmental Concerns Raised by the Rio Conference: The United Nations Conference on Environment and Development (UNCED) was held on 1992 at Rio de Janeiro is also known as the Rio Summit, Rio Conference or Earth Summit. It was a major United Nations conference, participated by 255 governments, with 144 sending their heads of state or government and 2,400 representatives of non-governmental organizations (NGOs) attended, with 17,000 people. An important achievement of this summit was an agreement on the 'Climate Change Convention' which in turn led to the 'Kyoto Protocol'. The agreement envisages that activities should not be taken on the lands of indigenous people, which would lead to environmental degradation or that would be culturally inappropriate. The Agenda 26.1 says that there is a close relationship between indigenous people and the natural environment. These communities are culturally, socially, economically and physically attached to forests and, therefore, promotion of sustainable development is highly essential at the national and international level, so as to recognize, accommodate, promote and strengthen the role of indigenous people and their communities (United Nations Sustainable Development).

Pressure from World Bank: The World Bank as a largest funding agency in the world as well as to India refused to give loans in the absence of a rehabilitation policy. In response, the first rehabilitation policy in India drafted by the Ministry of Rural Development in 1993 was the immediate result of the withdrawal of the World Bank from the Narmada project. Over the last decades, the development-induced displacement too has accelerated due to large scale investment of private capital for industrial and mining projects, primarily having been triggered by the process of liberalization. It resulted in stiff resistance and violence by the people for adequate resettlement and rehabilitation for the displaced. The Government of India realized the situation and problems and drafted the resettlement and rehabilitation policy in 1993 (Fernandes 1997: 36; Fernandes 1998: 2703; Fernandes 2004: 1191; Ministry of Rural Development 2004; Kujur 2005: 138–151; Asian Indigenous Tribal People Network 2006a).

In the absence of a uniform national policy on resettlement and rehabilitation, each ministry has drafted its own official guidelines to rehabilitate oustees affected by their respective development projects. On the national level, the Department of Irrigation under the Ministry of Energy and Irrigation had been the first to issue directives to State Governments to rehabilitate persons displaced by the major reservoir projects. On 19 May 1980, the Department of Irrigation gave the following directions to all state governments.

1. 'The forests and tribal area usually get submerged, but the benefits of the projects go to the rich farmers. Due consideration should be given to the aspect of submergence and compensation. Adequate provisions should be included in the project estimates for rehabilitation of such displaced as well as cash compensation for property loss'.
2. 'Displaced persons who lose land and property to be satisfied by the grant of wasteland in command area, or from large land holders'.

3. 'The State Government should ensure the protection of the weakest section of society before finally sanctioning the projects' (Pandey 1998: 10).

Our Constitution has made a certain provision and special priority for the upliftment of the tribals and for this purpose, the Ministry of Home Affairs issued a set of guidelines for the rehabilitation of displaced tribals, which was based on the recommendation made by a conference of Chief Ministers, held under the Union of Home Ministry on 17 February 1982. The recommendations are:

1. 'Tribals should not be evicted without being paid the full amount of compensation for the land acquired from them'.
2. 'Since the tribals are not quite aware of the correct market value of the lands they possess, a proper assessment of the acquired lands of the tribals should be done and the compensation money be deposited in banks in long term deposit scheme so as to benefit them. The project authorities should assist them in the sale and purchase of lands'.
3. 'Land based rehabilitation should be provided so that the tribal family is not completely uprooted from its traditional occupations'.
4. 'Emergency measures should not be undertaken while acquiring tribal lands and restriction should be made to check excess land acquisition'.
5. 'The State Government should enact laws to grant the displaced tribals land in the Command Area of an irrigation project'.
6. 'The project report for resettlement and rehabilitation should be prepared well in advance to ensure protection of the present socio-economic status of the oustees, and consider capacity of the project to absorb them'.
7. 'The displaced tribals should be rehabilitated not as individuals, but as a community, as far as practicable. Advance planning should be done and additional financial expenditure should be provided in the project report'.
8. 'To ensure livelihood, employment should be provided to at least one member of a tribal family in case of non-availability of land. If required, they should be

imparted proper training at project cost to acquire proficiency to get employment either in the main project itself or in other departments of the State Governments'.

9. 'Programmes of training in entrepreneurial skills, financial support, etc, should be provided for self-employment'.
10. 'Free house site should be provided to the affected families where houses of the oustees have been acquired'.
11. 'All civic amenities should be provided in the colonies set up for resettlement of the oustees' (ibid.).

A committee of Ministers was formed in 1982 to lay down norms for land acquisition and rehabilitation measures. It recommended that the displaced should be accommodated in the command area of the projects as far as possible. In such case, the large landholders may be compelled to part with a chunk of their land for the purpose of resettling displaced persons. Recommendation was also made relating to the issues like allotment of residential land, transport facilities, salvage material, ex-gratia payment, loan and advances, facilities and amenities in the new colony, employment on the project (ibid.: 11).

On 3 February 1986, the Bureau of Public Enterprises of the Department of Public Enterprises under the Ministry of Industry issued a set of instructions on land acquisition and rehabilitation issues involving major projects. Following these instructions, on 27 May 1986, the Ministry of Water Resources issued the following instructions to all the state governments for its implementation.

1. 'There should be a Rehabilitation Cell in each and every project to identify persons displayed due to acquisition of their land'.
2. 'Rehabilitation assistance should be given to the displaced persons for the acquisition of their land. Alternative homestead lands should also be provided for those who themselves cultivated those lands in their previous habitats. Absentee landlords will not be entitled to any rehabilitation assistance'.
3. 'A person losing his entire land will get priority over those losing partially. Persons losing homestead will be compensated by allotment of developed homesteads with all civic

amenities like roads, water supply, educational, medical facilities, etc. the cost of which will be borne by the project authorities'.

4. 'The project authorities should bear the cost of imparting suitable vocational training to the oustees according to their educational capabilities. If suitable training facilities are not available in existing institutes, such training courses, which would equip the candidates for employment, should be started and funded by the project. But the project authorities are not bound to provide employment after the completion of the training'.
5. 'There will be no assurance for employment opportunities in the project itself. But the project will assist the State Governments in organising and financing the oustees in vocational training programmes in poultry, animal husbandry, etc'.
6. 'The Rehabilitation Cell will monitor the progress of rehabilitation activities'.
7. 'In case of displaced tribals, the Ministry of Social Welfare and the Ministry of Environment and Forests should be associated with the preparation of their rehabilitation scheme' (ibid.).

In March 1984, a Committee was constituted by the Government of India to formulate policies for the rehabilitation of tribals, displaced of all types of development projects. It recommended a national policy covering all categories of displaced persons and incorporated three distinct provisions: (i) preparation of affected groups for transfer; (ii) transportation of the displaced to the new site; and (iii) integration of the displaced people with the new community who will be their neighbours in the site of their rehabilitation. Provisions were made to rehabilitate them near to the tribal areas on group/community basis, with free house site, water, concessional loans for supporting services like dairy, poultry, fishing, animal husbandry, forestry, carpentry, etc. Imparting training for skill enhancement and reservation of jobs should be made to the displaced tribals. It also recommended that special care should be taken for the food gatherers, hunter, forest produce collectors, forest land cultivators, shifting cultivators and tribal artisans depending upon forests raw materials.

Recommendation was made for involvement and participation of displaced tribals in the resettlement and rehabilitation programme for its effective monitoring and implementation (ibid.).

The Committee too suggested for the amendment of the Constitution to enable the State and Central Governments to enact laws on the subject. It recommended that there should be a full fledged Directorate of Rehabilitation in every State and State Level Rehabilitation Committee under the Chairmanship of Chief Secretary to review the programme and progress of resettlement and rehabilitation work (ibid.: 11–12).

Even though, in September 1987, 'National Water Policy' was framed, there were no specific guidelines for the resettlement and rehabilitation of persons displaced by irrigation projects. The policy, among other guidelines prescribed that:

> There should be an integrated and multidisciplinary approach to the planning, formulation, clearance and implementation of projects, including catchment treatment, environmental and ecological aspects, the rehabilitation of affected people and Command Area Development (ibid.: 12).

Besides official announcements and directives made by different governmental bodies, the Government of India also organized various regional and national seminars at different places of the country in order to understand the whole dynamics of water resource development, discussing its several socio-economic implications. The Water and Power Constancy Services (WAPCOS), a Government of India undertaking under the Ministry of Water Resources, conducted a number of seminars on 'Water Resources Development – Environmental Issues and Solutions' during the year 1987 on regional basis at Bhopal, Ahmedabad, Trivandrum and Bhubaneswar, highlighted the problems of rehabilitation and resettlement. A national seminar, organized in New Delhi on 29 February 1988, discussed the whole matter in detail and recommended the following:

1. 'A national policy for the resettlement and rehabilitation of ousted should be formulated and implemented effectively'.
2. 'In view of the special socio-economic status of a tribal people, as far as possible they should be ensured resettlement on a community basis'.

3. 'The national policy should immediately formulate suitable guidelines based on the decisions taken in a number of projects which may serve as common frame for resettlement and rehabilitation of the people already affected by the projects in different States' (ibid.).

Policy guidelines have also been made by the World Bank, the largest funding agency for Water Resource development projects in India. It has issued the following guidelines for involuntary displacement caused by such projects.

1. 'Displacement of Human Settlement should be avoided as far as possible. If unavoidable, it should be reduced to the minimum'.
2. 'All resettlement programme must be development-oriented so that the productive base and earning ability of the oustees are improved and they should share the benefits of the projects that displaces them'.
3. 'Socio-economic surveys of the affected area and of the population should be done well in advance'.
4. 'The Government concerned should take the responsibility of relocating the affected persons through an appropriate action plan'.
5. 'The affected people should be involved in planning as well as execution of the resettlement programme'.
6. 'The resettlement colonies should be provided with all kinds of civic amenities to ensure equitable treatment in the matter of development activities'.
7. 'The approach to resettlement programmes should be to ensure that the oustees regain their previous standard of living. As far as practicable, they should be integrated economically and socially with the host communities'.
8. 'A special organization should oversee the implementation of the rehabilitation and resettlement action plan through adequate provision'.
9. 'There should be timely monitoring of the progress and completion of the resettlement activities' (ibid.).

The first drafted policy of post-Independence India (*Draft National Policy for Resettlement and Rehabilitation of Persons Displaced as a Consequence of Acquisition of Land, 1994*) by the

Ministry of Rural Development too put moral obligations on other ministries to draft their respective resettlement and rehabilitation policies to address the problems of the displaced. Such are: *Draft national Policy for Resettlement and Rehabilitation of Persons Affected by Reservoir Projects, 1994* by the Ministry of Water Resources; *Resettlement and Rehabilitation Policy 1993* by National Thermal Power Corporation; *Resettlement and Rehabilitation Policy* by Coal India Limited, etc. (ibid.: 9).

The Constitution of India in its Seventh Schedule provides for the relief and rehabilitation of persons displaced from their original place of residence by reason of the setting up of the Dominions of India and Pakistan. Accordingly, the Government of India evolved certain policy guidelines for the resettlement of persons displaced from Pakistan, Bangladesh and other neighbouring countries due to political reasons. But there is no comprehensive policy for the resettlement of those who have been displaced due to the establishment of various development projects within India. It is true that the Constitution, in its Fundamental Rights chapter, does entitle citizens to reside and settle in any part of the territory in India [Article19 (1)(e)], but at the same time this right is almost nullified by the *sub-clause subject to reasonable restrictions in the interest of general public.* As a result, even the special privileges granted to the Scheduled Tribes can get curtailed (ibid.: 9–10).

The Government of India in course of time realized the loopholes of existing policy and further modified the R&R Policy from time to time. The policy of 1993 drafted by Ministry of Rural Development was modified in 1994 after the consultation with the 15 ministries and departments. The second draft of 1994 can be said to reflect the thinking of the government of India as a whole while the first is that of the Ministry of Rural Development. Hundreds of social activists and thousands of displaced persons prepared an alternative to its second draft of 1994 and presented it to the secretary, Ministry of Rural Development on 5 October 1995. The Ministry of Rural Area and Employment prepared a new draft and it was approved by the committee of secretaries on 28 November 1997. Although it brought certain changes in the policy, there were some shortcomings. The 'public purpose' for which land is acquired under Land Acquisition Act has not been defined properly. In the absence of a definition, courts have generally accepted as 'public purpose' whatever the government declares including the transfer of land to the private sector. The social activists and the displaced people has been demanding a restrictive definition and welfare of the biggest possible number, such as educational and health institutions and housing for the poor (Fernandes 1998: 2703; Fernandes 2004: 1191).

In 1998, another draft came out along with the amendments to the Land Acquisition Act 1894. The National Policies on Resettlement and Rehabilitation (NPRR) 1998 of course made some changes; it does not give adequate importance to the cultural aspect. The gender issue does not receive the importance it deserves. Though rehabilitation was declared as mandatory, it is not mentioned as fundamental rights. It remains as welfare measures. The rehabilitation of the displaced people in the past has not been mentioned. According to the estimates, at least 213 lakh people were deprived of their livelihood between 1951 and 1990 and among them 40 per cent are tribals. Significantly the policy is not binding as long as it becomes law and NPRR 1998 has no provision for it (ibid.).

The heavy resentment over the NPRR of 1998 among the displaced people and the social activists again forced to prepare a draft National Policy on Resettlement and Rehabilitation. The Ministry of Rural Development finalized and circulated the NPRR 2003 and notified on 17 February 2004. The NPRR of 2003 provided the displaced some special benefits. It includes the special assistance to the non-land holder or the daily wage labourers at least 500 days of minimum agricultural wages for the loss of customary/grazing/fishing rights and other R&R benefits. The monetary compensation for the loss of assets extended up to 25 per cent and they were given the fishing rights in the river as well as in the reservoir. In terms of resettlement, provision was made to resettle them in a compact bloc according to their ethnic, linguistic and cultural identity. But there were lots of difficulties in this policy. The people (landless agricultural workers, forests dwellers, artisans, etc.) who are critically dependent on forests land for their livelihood were not eligible for cash compensation. The people who have their legal document or *Pattas* will be entitled for compensation. This policy will be applicable only to projects displacing 500 families in the plain areas and 250 families in the hilly areas. Further the policy did not give any provision for the past oustees and was silent about environmental protection and community participation. The biggest lacuna in both the policies of 1998 and 2003 is absence of provision for the Partially Affected Persons and the displaced communities do not have any participation in the decision-making process in the course of their own future (Ministry of Rural Development 2004; Kujur 2005: 138–151).

The loopholes of the NPRR of 2003 and its severe criticism prompted the government to re-evaluate it. The Government of India drafted another policy in 2006 known as Draft National Rehabilitation Policy (NRP) 2006. Its preamble recognizes traumatic, psychological and socio-cultural consequences on the displaced populations which call for affirmative state action for protecting their rights. It has empowered the affected people to participate in the process of 'Social Impact Assessment', Environmental Impact Assessment', 'desirability and justifiability of each project', and the 'Tribal Development Plan' etc. NRP 2006 promises all happiness for displaced persons could desire for. But the 'Preamble' is necessarily not the actual part of the Policy and, hence, does not constitute safeguard. First, the affected persons do not have the right to be consulted prior to finalization of their lands as the project site. Secondly the people have no right to participate or represent while preparing of Social Impact Assessment (SIA) report and Environmental Impact Assessment (EIA) report. The Policy provides for constitution of a 'multidisciplinary expert group' to examine the SIA and EIA reports. Members are nominated by central and state governments. There is no accountability, independence and transparency in the process of examination of the SIA and EIA reports. The NRP–2006 gives sweeping powers to the Ministry of Defence to acquire land in connection with national security and the Ministry is exempted from conducting any

Social Impact Assessment or Environmental Impact Assessment. The Policy fails to provide adequate safeguards to the displaced persons. Only the project affected families who are/were having possession of forest lands prior to the 25 October 1980 will be entitled to the benefits of the NRP–2006. The Policy neither guarantees land for land nor adequate compensations to the affected families. Affected families shall get a one-time financial assistance of only ₹ 10,000 per hectare for land development and get maximum of ₹ 20,000. In case of allotment of agricultural land, one-time financial assistance of only ₹ 5,000 will be provided. There is little guarantee for employment for the oustees. Employment will be according to the availability of vacancies and suitability of the affected person for the employment. There is absolutely no rehabilitation for partially affected families. Tribals who have been residing continuously for a period of three years in the displaced area would come under the strict definition of 'affected family' to get the benefits under this Policy. The people who practice traditional mode of agriculture, such as shifting cultivation or temporarily shifting from one place to another place every year for cultivation of crops, and other nomadic forms of life would not get compensation (Asian Indigenous Tribal People Network 2006a).

The National Rehabilitation Policy of 2006 was further re-formulated in 2007 known as the National Rehabilitation and Resettlement Policy 2007 notified on 31 October 2007. The 2007 Policy was supposed to be an improvement of the Draft National Rehabilitation Policy of 2006. There has been a positive improvement in the 2007 Policy as it provides prior consultations with the concerned Village Councils in the sixth Scheduled Areas as well as with concerned Gram Sabhas (Village Council) in the fifth Scheduled Areas in all cases of land acquisitions including land acquisition in cases of urgency under Land Acquisition Act of 1894. Earlier, the 2006 Draft Policy had such provision only for the land acquisitions in the fifth Scheduled Areas. But the 2007 Policy does not provide guarantee for land-for-land compensation which is mandatory under the Constitution for the Scheduled Tribes living in the fifth and sixth Scheduled Areas (ACHR Weekly Review 2007).

The 2007 Policy still upholds the sovereign power of the State to apply the concept of *eminent domain* to forcibly acquire any private property in any part of the country in the name of *public purpose*. This power is provided under the Land Acquisition Act of 1894. The 2007 Policy deletes previous provisions of the 2006 Draft Policy which provided that emergency provisions under Section 17 of the Land Acquisition Act of 1894 should be *used rarely* and should be applied only after considering *full justification* of the proposed project (Clause 6.23 of the 2006 Draft Policy). Clause 7.18 of the 2007 Policy implies that land can be acquired in case of emergency under Section 17 of the Land Acquisition Act, 1894 or similar provision of any other Act of the Union or a State for the time being in force by keeping the affected families in *transit and temporary accommodation, pending rehabilitation and resettlement scheme or plan* (ibid.).

Clause 4.7 of the 2007 Policy further exempts the Ministry of Defence from conducting any Social Impact Assessment or Environmental Impact Assessment while acquiring any land in connection with national security. So, for example, if a nuclear plant were set up for national interest or defense purposes or in the case of border fencing, no one can oppose the project (ibid.).

As India is a federal state, both centre and the states have their own resettlement and rehabilitation policy. Maharashtra was the first state in the country to pass a law in 1976. Madhya Pradesh passed a law in 1985. The Karnataka Act was passed by the legislature in 1987 but received the president assent in 1994. Orissa had passed several Government Orders (GOs) beginning from 1977 and finally a resettlement and rehabilitation policy for irrigation displaced persons was promulgated for the first time in 1994. Many of the states do not have their own resettlement and rehabilitation policy after 60 years of India's Independence. Only few of the Indian states have promulgated their own Resettlement and Rehabilitation Policy (Fernandes and Paranjpya 1997: 16).

The Right to Fair Compensation and Transparency in Land Acquisition, Resettlement and Rehabilitation Act, 2013

Raising the issue of involuntary displacement and its large scale socio-economic impacts, the civil society activists, academics and non-government organizations have long been demanding the scrapping of the colonial law – Land Acquisition Act 1894. They says that the law is undemocratic and inhumane, which doesn't protects the interests of the affected persons. They have been demanding certain changes in the existing law, which are extremely detrimental to the affected persons. Hence, they advocated for enactment of new law incorporating these concerns like consent of the likely displaced persons while acquiring land; restricting the ambit of 'public purpose' to only those activities undertaken by the government which directly benefit the people, particularly the likely displaced; ban on acquisition of land for private companies; mandatory environmental and social impact assessment before taking a decision to acquire land; prohibition of acquisition of irrigated and multi-cropped land; obligation to provide compensation, rehabilitation and resettlement to all the displaced affected by acquisition of land either by state or private entities and return of unutilized land acquired in excess of the requirement of the project to erstwhile owners (Saxena 2015: 324–336).

The mounting pressure of the protests and movements across the country delayed the execution of number of high profile projects and it forced the government to undertake complete revision of the colonial Land Acquisition Act 1894. However, this is not say that it happened only in one stroke, the exercise was started earlier; the first draft was drafted in the year 2007 and it followed the second draft in 2009 and the final draft was enacted in the year 2013. The most awaited act came into to force after much deliberation in the parliament, consented by both the ruling United Progressive Alliance (UPA) and the opposition parties, National Democratic Alliance (NDA). The new act is known as the Right to Fair Compensation and Transparency in Land Acquisition, Resettlement and Rehabilitation Act 2013. But the irony is that, the new act, before its implementation started, the NDA government which came to power in the year 2014 decided to take ordinance route to amend the most important provisions of the act, which is trying to restoring the durability of the colonial Land Acquisition Act, 1894. The amendment has added a new chapter III A, which empowers the government to exempt from the following five categories of projects from the obligation of obtaining consent of the families affected by acquisition of land: (i) project vital to national security of defence of India and every part thereof, including preparation for defence of defence production; (ii) rural infrastructure including electrification; (iii) affordable housing and housing for the people; (iv) industrial corridors and (v) infrastructure and social infrastructure projects including those of public and private partnership (Saxena 2015: 324–336; Government of India 2013: 1–45, Government of India 2014: 1–4).

Debate and discussion has been going on that, the present government in the exercise of amending the new act of 2013 has been triggered by the pressure of corporations, bureaucrats, politicians and neo-liberal economists, who considered the act of 2013 is unworkable and it act as a serious bottleneck to the economic growth of the country (Saxena 2015: 324–336). However, the present NDA government's Land Acquisition Amendment Bill in the Lok Sabha on 10 March 2015 has been resisted by key political parties. The proposed amendment is called anti-farmer and anti-poor. While presenting the bill in on the floor of the parliament, the Bill was passed in the Lok Sabha, since the ruling government has majority in that house, but it still needs approval from the Rajya Sabha, where the government doesn't have a majority to pass the bill. However, the government has now agreed to bring back the key provisions of the Right to Fair Compensation and Transparency in Land Acquisition, Resettlement and Rehabilitation Act 2013, which was triggered by the mounting pressure of the opposition political parties. The government retracting from its previous position, there is likelihood that the panel will come out with a consent report soon (Biju 2015).

Resettlement and Rehabilitation Policy in Orissa

The State of Orissa occupies a unique place among the underdeveloped states of India because of its larger concentration of tribals population. The most striking feature of Orissa is that it is rich in resources, but people are extremely poor. It is endowed with minerals like coal, iron ore, bauxite, chromites, manganese ore and graphite etc. Being a land of rivers, it has vast potential of water resources for irrigation and hydropower generation. Yet, it remains the most drought prone state in the country. The bulk of this potential of water resources is utilized by the multi-purpose or major and medium dam projects. At the same time, the availability of rich minerals of different varieties in the region has also encouraged the growth of large-scale industrial and mining projects in the state. Although development activities in Orissa began in the late 1940s, it gained real momentum during the post-Independence era with the introduction of planned development (Pandey 1998: 31–32).

The notable projects in the 1950s were in the Rourkela Steel Plant and Hirakud Multi-Purpose Dam. The Hindustan Aeronautics Limited (HAL), Talcher Thermal Power Station (TTPS), and the Balimela dam came in the 1960s. The 1970s saw the establishment of Rengali, Upper Kolab, Upper Indravati and Subarnarekha multi-purpose dams. The Ib Thermal Power Station (ITPS), Talcher Super Therma Power Project and the National Aluminium Company (NALCO) were set up in the 1980s. These projects were executed in resource-rich regions, which had been inhibited for generations by the tribals and the rural poor. Though these development projects indeed brought manifold benefits to the states, on the flip side, it caused large scale involuntary displacement. The groups displaced were mostly the weaker sections of society belonging to the Scheduled Castes, Scheduled Tribes and other backward classes (ibid.).

Looking at the resettlement and rehabilitation policy for the displaced in Orissa, it has passed through more or less the same stages as the rest in India (Jena 1998: 821). There was no policy of resettlement and rehabilitation up to 1973 for the displaced persons. The government of Orissa was following a mere compensation package for acquired land according to the 1948 Orissa Act. The resettlement issue raised its peak in 1971 when Rengali dam project saw a strong people reaction. In 1973, Orissa came up with the first resettlement policy and in 1977 accepted land for land as the basis for rehabilitation. Later in 1994, the Department of Water Resources for the first time formulated a comprehensive progressive policy on resettlement and rehabilitation for all water resources project. The stiff resistance and violence by people for adequate resettlement and rehabilitation further forced the Government of Orissa to promulgate the R&R

Policy in 2006, which is the improvised policy of the 2003 National Policy on Resettlement and Rehabilitation (Jena 1998: 821; Jena 2006: 385; Fernandes and Paranjpya 1997: 34–35).

Orissa did not have a well formulated rehabilitation policy before Independence. This is quite clear from the way the rehabilitation aspect was treated in the case of Inter-State Machhkund Hydel Project, which was undertaken in the early 1940s. Soon after Independence, in 1948, the Government of Orissa adopted the *Orissa Development of Industries, Irrigation, Agriculture, Capital Construction and Resettlement of Displaced Persons (Land Acquisition) Act*. The Act only guaranteed compensation, which had to be paid by the competent authority to the persons. Nothing was mentioned regarding the resettlement of the displaced under the Land Acquisition Act. Since there was no uniform national policy on resettlement and rehabilitation of oustees in Orissa, rehabilitation packages were provided by different government departments, public sector and private sector agencies differed from one another. Even in the case of irrigation projects, there was no uniform policy till August 1990. Surprisingly, the public sector (central) undertaking, National Aluminium Company, follows two different sets of rehabilitation policy in its two separate sub-projects (Pandey 1998: 12–13).

In the post-Independence India, Hirakud dam is the first major multi-purpose project after planned development took off in the country. It was followed by the Salindi irrigation project and the Balimela multi-purpose reservoir projects. It caused large scale displacements resulted from all those projects and they remain contended whatever compensation they received for their land and property. Historically speaking, the problem of displacement of people and their subsequent rehabilitation gained importance in Orissa, when the construction of Hirakud dam was constructed just after the Independence. There were no such special guidelines for the displaced people and, thus, ad hoc policy was adopted for the Hirakud dam. However, it created several problems for many among the oustees. There were many shortcomings of this ad hoc policy and it caused many of the displaced people as landless and homeless (ibid.: 13–14).

However, with the commencement of the Rengali dam project in the 1970s, a new chapter in the history of resettlement and rehabilitation in Orissa began. The political awareness and growing demands of legitimate rights by the people in the 1970s brought several agitations to have their proper rehabilitation. This forced the Government of Orissa to evolve a resettlement and rehabilitation policy in 1973, and which was formulated for the Rengali project. Although agitations during the 1950s and 1960s continued for two decades, the resistances of the displaced were not well organized and long-standing. However, it strengthened during the 1970s and the Government of Orissa considered for resettlement and rehabilitation especially for the Rengali dam project. The World Bank too issued guidelines for the resettlement and rehabilitation of affected people by the projects funded by the World Bank. The rehabilitation made for the benefit of the affected people by the Rengali dam project was further extended to Upper Kolab and other medium projects after some amendments, which were under construction by that time. Further, a revised resettlement and rehabilitation policy was announced on 20 April 1977 and it was an improvement on the earlier one adopted for Hirakud and Balimela dams projects. The two main features of this policy are:

1. Lands for resettlement in government sponsored colonies were to be allotted to the ousted irrespective of the fact whether the oustees were landless or not;
2. Displaced persons who wanted to make their own rehabilitation arrangements in lieu of the land that was acquired from them were to be given the cash that might they require to by the land that was due to them (ibid.: 15–16).

A number of amendments were made later to the 1977 resettlement and rehabilitation policy in view of the multi-faceted problems faced by the oustees to build their homes, restore their livelihoods and cope with the new environment, etc. Following these problems, a high level meeting was held on 8 June 1990, presided over by the Chief Minister of Orissa. As a result, for the first time a uniform resettlement and rehabilitation policy for irrigation projects came into being on 21 August 1990. Between 1990 and 1993, several substantial amendments were made to the policy which brought about further improvement with the active involvement of Non-Governmental Organizations (NGOs) and the World Bank to extend more benefits to the project affected persons. As a result, on 27 August 1994, the Department of Water Resources, Government of Orissa, came up with the Orissa Resettlement and Rehabilitation policy, which is the revised version of the Resettlement and Rehabilitation Policy of 1990 (Pandey 1998: 13–19; Jena 1998: 821; Jena 2006: 385; Fernandes and Paranjpya 1997: 34–35).

This policy is considered to be more liberal compared to the resettlement and rehabilitation policies of the various projects in the state. It widens the definition of Displaced Peoples (DPs). It includes those people, whose means of livelihood are affected directly by the result of acquisition of land. A village will be declared fully submerged if more than 75 per cent of its agricultural land is submerged. It also recognized women rights, widows, divorcees, orphans and physically challenged. An important feature of the policy was to provide identity cards for the displaced and resettle the tribals in groups near to the project area, those who have their customary rights over the forests products. More or less, the policy of 1994 had certain benefits for the displaced, but some of the provisions were not acceptable. The basic assumption of the policy maintained displacement as inevitable. It does not speak of rehabilitation as a right. It continues to be a welfare measure. The policy was limited to only water resource projects and not all kind of development induced displacement. In terms of cash compensation it was applicable to only *Patta* holder. The tribals who have their customary rights over the forests produce were debarred from such compensation. There was no provision of compensation for the trees. The term 'public purpose' had not been defined. Thus, the provision of eminent domain continues to prevail. Moreover, there was no indication that it will become a law. Finally, it has been argued that the policy was an effort to get funds from the World Bank and stop the agitation of the displaced people by giving them some sops (Fernandes and Paranjpya 1997: 34–35; Mahapatra 1995: 318–339).

In the post-liberalisation of Indian economy, Orissa is the first to have attracted the highest foreign direct investment. In

2005, the Government of Orissa has already signed 43 Memoranda of Understanding (MoUs) and it opened the flood-gates of large scale industrialization and mining in Orissa. The massive displacement of people caused by such projects led various protests and movements across the state. The Kalinga Nagar incident on 2 January 2006 is part of a series of resistance of the people threatened with displacement by mega development projects in the state. It took death of 12 tribal protesting against the establishment of a plant by the Tata Iron and Steel Co Ltd on their lands. The Kalinga Nagar tragedy is an indication of the nature of implementation and the impact of development induced involuntary displacement in Orissa. It assumed a greater urgency and imperative for the Government of Orissa to adopt a pro-people, pro-displaced and begin a Resettlement and Rehabilitation Policy to continue the process of industrialization. Just after four days of the Kalinga Nagar incident, on 6 January 2006, the government constituted a ministerial committee to finalize the Resettlement and Rehabilitation Policy for the affected communities known as Orissa Resettlement and Rehabilitation Policy (ORRP), 2006. The policy is applicable to all development projects in public, private and joint sectors. It covers displacement resulting from dams, canals, flood control works, power stations, mining, urban housing and shopping complexes, slums clearance, roads, railways, airports, seaports, conservation, parks/bio-reserves/sanctuaries, sport complexes, amusement parks and defense establishments. In comparison to Resettlement and Rehabilitation Policy of 1994, the Policy of 2006 has extended its scope. The policy of 1994 was only dealing with the displaced by water resource projects; where as the policy of 2006 deals with all kinds of displacements (Jena 2006: 384; Asian Indigenous Tribal People Network 2006c; Government of Orissa 2006a).

The ORRP provides for employment to an eligible member from each displaced family in cases of displacement by Industrial and Mining projects. It provides one time cash assistance up to maximum of ₹ 5 lakhs for displaced families who have lost all land including homestead land by industrial and mining projects. Provision for granting of free homestead land of 1/10th acre in the resettlement habitat will be provided according to the availability. If land is not available, it provides for compensation of ₹ 50,000. The ORPP also provides for house building assistance of ₹ 1.5 lakh for each displaced family and a monthly maintenance allowance of ₹ 2,000 to each displaced family for a period of one year (Asian Indigenous Tribal People Network 2006b).

However, the ORRP 2006 cannot address the root causes of the dispossession of the *adivasis* in the state. The government of Orissa has been acting as broker and profiting from the lands of the peoples by exercising its sovereign power. In 2004, the state government had sold 2,000 acres of land to the Tatas for ₹ 3.35 lakhs per acre. But the state government paid only ₹ 76,000 an acre to the original landowners, where as the current market rate for an acre was ₹ 5 lakhs. It was applicable only for titleholders, who account for 20 per cent of the tribals in the area. The promise of a job for each displaced family, a home allowance of ₹ 50,000 and ₹ 5,000 as aid for a temporary shelter were never adequately fulfilled. The ORRP 2006 is meant only for families who would be displaced in future and it has not dealt with the 1.4 million already displaced people in the past, mostly indigenous peoples. It does not provide for free, prior and informed consent of the affected families before acquisition of the land. There is no mandatory provision for land as rehabilitation benefit. Only in case of displacement by irrigation projects, national parks and sanctuaries the displaced would get land as rehabilitation benefits. Since the majority of the displaced comprised indigenous and tribal peoples who primarily depend on agriculture and forest related activities, their livelihood would be adversely affected in the absence of land as a rehabilitation benefit. This violates provisions of the fifth Schedule of the Constitution of India which guarantees the rights of the tribals over their land and prohibits transfer of tribal lands to non-tribals. In the historic Samata Judgment of July 1997, the Supreme Court of India held that all the lands leased to private companies by the government of Andhra Pradesh are unconstitutional and, hence, null and void. The Policy too empowered for effective participation of the displaced communities in the Resettlement and Rehabilitation process, but there is no clearly defined mechanism for it (ibid.).

4

Dams, Controversies and Conflicts

In recent years, there has been a growing scarcity of water and it has become a favoured subject for seminars and conferences across the world. The UNDP, the World Bank, the Asian Development Bank are seriously concerned about the projected water scarcity. Numbers of academic institutions in many countries are engaged in research programmes and the possibilities of conflicts over the scarce natural resources, particularly water. Currently, there is a fashionable thesis that future wars will be fought over water, not for oil. Number of institutions and networks have sprung up to deal with this and related matters: World Water Commission, World Water Council, Global Water Partnership and so on. Different countries have taken the exercise of 'Water Vision 2025' under the aegis of the Global Water Partnership. The 'Vision' exercises were partly national (India Water Vision, Pakistan Water Vision, etc.) and partly thematic (Water for Food, Water for Nature, etc.). Eventually, these were brought into a 'South Asia Water Vision 2025' for presentation in the World Water Forum, held at Hague in March 2000. In India, such consciousness led to the establishment of National Commission on Integrated Water Resource Development Plan. The National Commission on Water finished its work, submitted its report in 1999, and made numerous recommendations (Iyer 2001: 1115).

Water becomes political wherever and whenever it is a scarce and valued resource. Water is important for the survival of human beings; be it for the purpose of domestic, agriculture, industry, navigation, fishing and tourism, etc. In most countries, water is scarce and becoming scarcer and its politics is taking on serious dimensions even in developed countries like the United States of America (USA) and Canada. In case of former, water is desperately fought over to sustain cities and commercial agriculture, and in case of the latter, it is widely seen as a diminishing resource. But in the Third World countries, water has always been a scarce commodity, which has been exerting pressures on existing water supply (Wood 2007: 36).

Nature of Water Conflicts

In India, water scarcity has become a common phenomenon and has been causing several types of water conflicts. First, water conflicts occur between or among water users who are dependent on the same source of supply or when usage increases and/or the resource diminishes. At the macro level, the source of water is usually a river or lake and at the micro level it might be a stream, a tank, a groundwater aquifer, or a tap at a village well. Conflicts begin when water supply depletes either on account of natural reasons (for example, lack of rain), or improper maintenance of resource structure (for example, a canal silts up) or environmental degradation (deforestation causes increasing aridity and a reduction in the water table). Water scarcity and conflicts too occur, while there is increase in number of users, or when they increase their quantity of consumption (ibid.: 37).

Second, water conflict arises when planning for a new development project (dam) on a yet-untapped water source opens up. It raises the question of costs and benefits of such project. On the one hand, it benefits few (in terms of flood control, irrigation, hydropower production, navigation, water supply etc) and, on the other hand, it puts adverse impacts on others (in terms of human displacement, their loss of livelihoods and environmental degradation) (ibid.).

Third, where the water source is already being tapped, water conflicts occur when there is an actual or anticipated change in the existing hierarchy of users even if supply remains constant. This basically means water users, those who previously received more water from the source, anticipate that they are or will be getting less, while those who previously received less now perceive that they can get more. This type of conflict occurs while using river water in the context of upstream and downstream. Conflict arises when farmers at upstream draw more water and it adversely affects farmers at the downstream (ibid.: 37–38).

A fourth source of conflict occurs when large dams are made on rivers and it becomes much more difficult to compute the costs and benefits of such projects. Large scale displacement, loss of habitat, source of livelihood, environmental degradation, loss of sacred places such as temples, pilgrimage, graveyards, etc. led to protests and movements against such projects. Thus, conflicts erupt over projects between 'government and pro-developmental forces' and 'victims of development' (ibid.: 38).

Fifth, conflicts too occur when government officials, usually of the Ministry of Water Resources, including top officers as well as local irrigation engineers, plays partial role. At the level of the national government it may be perceived that top ministerial officials favour the claims of one state in a river water dispute. At the micro level, frustrations build when officials are perceived to be partial in their management of the resource, usually for reasons of corruption (ibid.).

Sixth, sometimes conflicts are motivated by demands of specific interests or calculations of electoral gain. In a democracy like India, politicians represent the interests of their constituents if they want to get (re-)elected and it becomes a major element in disputes. Particular groups like large farmers, industrialists, fishing communities, or citizens of a municipality or village, organized themselves in a better way and put pressure on decision-makers (ibid.: 39).

Water Resource, Development and Administration

Over the past 50 years, India has been facing the problem of water crisis and it has become a politicized resource. Although India receives good rainfall, it is very uneven and it causes floods in some part of its territory and droughts on other parts. India shares 16 per cent of the World's population, 2.45 per cent of land resources and 4 per cent of its water resources. As far as rainfall is concerned, there are wide variations, both temporal and spatial. Much of the rainfall occurs within a period of few months during the year and even during that period, the intensity is concentrated within a few weeks. Spatially, there is a wide range in precipitation – from 100 mm in Rajasthan to 11,000 mm in Cherrapunji. Although it is known as among the wettest places on the earth, it suffers from an acute shortage of water in the dry month, because all the rain that falls quickly runs off the area. Sixty per cent of the water resources of India are to be found in the Ganga, Brahmaputra and Meghna river systems, which account for 33 per cent of the geographical area of the country; 11 per cent in the west-flowing rivers south of Tapi covering 3 per cent of the area; and the rest 29 per cent in the remaining river systems spread over 64 per cent of the land area. Broadly speaking, Himalayan Rivers are snow fed and perennial, where as peninsular rivers are dependent on the monsoons and, thus, these are seasonal. Again, the north and east are well endowed with water, whereas the west and south are water-short. Apart from the desert areas of Rajasthan, there are arid or drought-prone areas in parts of Gujarat, Maharashtra, Karnataka, Andhra Pradesh and Tamil Nadu. On the other hand, most of the time, the eastern part of the country experiences devastating floods from time to time (Iyer 2001: 1116–1117).

Responding to these temporal and spatial variations of rain fall and the crisis of water, our planners found the answer in constructing 'large storage reservoirs' and 'inter-linking of rivers' to mitigate such problems. However, construction of dams (major, medium and small) received greater attention and acceptability and, therefore, almost all rivers in India are dammed today, which was much appreciated for its multi-purpose functions of hydropower production, irrigation, navigation, flood management, water supply and tourism, etc (ibid.: 1117). The engineering intervention and the activities of damming rivers caused conflicts in many cases. Because many of those flows are inter-state in nature. All major rivers, including the Indus, Ganga, Yamuna, Brahmaputra, Narmada, Mahanadi, Godavari, Krishna and Cauvery, are inter-state rivers insofar as they originate in one state and then flow through another state or states before reaching the sea. Several smaller rivers, including the Damodar, Mahi, Tapi and Pennar, are also inter-state rivers (Wood 2007: 59).

During the last 50 years, most of these rivers have been the subject of disputes among co-riparian states over the allocation of the costs and benefits of developing river water resources. Costs include the cost of dam and canal construction and also that of displacement and damage inflicted by the creation of reservoirs and canal systems. Benefits include allocation of water for domestic use, drinking water, irrigation, industry, hydropower, flood control, fishing, navigation or tourism, etc. Most of the disputes have been of the 'upstream-downstream' variety (ibid.).

In order to understand the nature of water disputes in contemporary India, it is useful to look back at the kind of irrigation administrative structure that developed before Independence. During the British period, irrigation was managed by the military engineers of the East India Company. The Military Board, which supervised all irrigation works, was replaced in 1854 by the Public Works Department (PWD). However, Military engineers continued to dominate until 1895 and after which the PWD became purely civilian. In the year 1866, Government of India centralized the control of irrigation projects in its own hands, giving less regards to British provinces or princely states. All projects had to be approved by the Secretary of State. In case of disagreement, the Secretary of State's decision was final (ibid.: 60).

With the introduction of the Montagu-Chelmsford reforms in 1919, irrigation became a provincial but 'reserved' subject. Provincial legislatures could discuss projects and even raise loans to finance them, but control over expenditure and administration was reserved to the governor in council and, therefore, fell under the centralized control of the Secretary of State. Whenever more than one province or state was involved prior approval for all aspects of a project had to be obtained from the Secretary of State. However, with the promulgation of the Government of India Act of 1935, irrigation became one of the subjects where centralized imperial control was loosened and central government's legal competence in respect of irrigation and water was confined to inter-state disputes. On the whole, the dispute-settlement system was deemed to work reasonably well, and the combination of state jurisdiction over water plus central arbitration in the case of disputes was carried over into the Constitution and laws of independent India (ibid.: 60–61).

Soon after the Independence, major changes took place and it completely altered the context of inter-state river disputes and the politics of water in India. The process of democratization and people's demands of basic necessities of life like food, shelter, clothing, health care, and especially water brought large scale changes. The bureaucracy trained by the British to keep order and collect revenue was gradually remoulded to deliver programmes that would improve productivity and social justice in every aspect of development, from agriculture and health to literacy and village industries. The First Prime Minister of India, Jawaharlal Nehru's development philosophy was rooted in a belief in the ability of science and technology to solve human problems (ibid.: 61–62).

Water as a subject and its administration was changed during the post-Independence in the context of decentralized federation along with reorganization of states on a linguistic basis. From the beginning, decentralization was compromised by a fear, born out of the experience of Partition in 1947, that disunity and separatism might result if too much control was given away by New Delhi. The tendency against decentralization was reinforced by Nehru, which gave immense powers to a central Planning Commission and made the states dependent on the Union government's control of major revenue sources. Although, jurisdiction over many aspects of India's rural development was given to the states by the Constituent Assembly, but in reality the Central government retained a great deal of developmental initiative and control. Water-resource development is a case in point. Thus, the centre-state political relationship became exceedingly complex in matters of inter-state river water development (Wood 2007: 61–62; Chokkakula 2012: 6–12).

The Constitution of India in its Seventh Schedule says 'Water, that is to say, water supplies, irrigation and canals, drainage and embankments and water storage' falls under the state jurisdiction. However, it has also kept the provisions in the entry 56 of List I, which says that central government has the authority to control, regulate and development of inter-state rivers and valleys for the sake of public interest. Article 262 of the Constitution empowers the central government to provide for the adjudication of 'any dispute or complaint with respect to the use, distribution or control of the waters of, or in, any inter-state river or river valley.' As per the above provision, the Parliament enacted Inter-State Water Disputes Act, 1956 and the River Boards Act, 1956 (ibid.).

The River Board Act, 1956, enabled the central government to create, in consultation with the relevant state governments, boards to advice on the integrated development of inter-state basins. The basic purpose of the act is that it would facilitate the optimum use of river water and coordinate the development of irrigation, drainage, canals, flood control and hydro-electric power. The river boards were supposed to prevent disputes by preparing developmental schemes and working out the costs to each state. Unfortunately, to date no river board has ever been created and the River Boards Act has been termed as 'dead letter' by the Sarkaria Commission on Centre-State Relations.

Inter-State River Disputes

There are number of cases of inter-state river disputes in India and a few among those are discussed here.

Damodar Valley Corporation: The Damodar is an inter-state river and it is the India's first post-Independence river valley project and it was supposed to serve as a model for future river water resource development. However, it became conflictual. The problem was not of water scarcity, but control of floods. The Damodar originates in the hilly Chota Nagpur region of Jharkhand (earlier part of Bihar) and it flows southeast into West Bengal where it is joined by its main tributary, the Barakar. Near Burdwan, the Damodar turns south and eventually joins the Hooghly River about 120 km (74 miles) south of Kolkata. Inspired from the Tennessee Valley Authority in the US, planning was started in India to replicate and implement the model in Damodar for it multi-purpose benefits of irrigation, hydropower and flood control, etc. The Damodar Valley Corporation Act was passed in the Lok Sabha on 18 February 1948 (ibid.: 64–66).

Although water is a state subject, the central government took the initiative in creating the Damodar scheme. This is largely due to the inability of West Bengal and Bihar to cope with such a large project in the early years of Independence and disruption made on Partition of the country. Large scale projects were executed under central planning and Nehruvian state-directed economy. The Damodar Valley Corporation (DVC) was intended to have a great deal of autonomy, but it became subject to increasing political pressure from the participating governments (ibid.).

By 1958, four out of eight planned dams had been built on the Damodar at Konar, Maithon, Panchet Hill and Tilaiya. But even by 1953, West Bengal was withdrawing its support for the development of the remaining four dams. Instead, it went into competition with DVC by building its own thermal power units. Meanwhile, Bihar pressed for building more dams because it needed their hydroelectric output to power the industrial expansion that was occurring in its part of the river valley, particularly the large steel plant at Bokaro (ibid.).

With regard to irrigation, the DVC imposed a 'betterment levy' to extract payment for water from farmers and also to pay the cost of dam construction. But West Bengal refused to pay and many West Bengal farmers began stealing water from the canals. The opposition parties in the West Bengal Legislative Assembly began a campaign to oppose not only the levy but also other irrigation fees to be imposed by the state. Even, the West Bengal government refused to collect the levy and allowed farmers to sabotage irrigation channels and failed to pay its share to the DVC or repay its debt to the central government. The final blow to DVC came in the year 1959, when one-third area of West Bengal was affected by floods. Over 100 people lost their lives and 2.3 million were on flood relief. There was immense damage of crops, livestock, communications, transport and industry. West Bengal politicians argued that the floods were caused by DVC negligence. On the other hand, Bihar argued that the flooding could have been lessened had the remaining dams of the DVC been built. The most significant aspect of the dispute is that the central government took no action despite reports from two commissions of its own investigative engineers that supported Bihar's position. Today, DVC has ceased to function as a river valley authority and simply operates as a power-generating body (ibid.).

Krishna and Godavari River Disputes: Godavari is the largest, and Krishna the second largest of the east flowing rivers of peninsular India. Both rivers originate in Maharashtra and flow through Karnataka and Andhra Pradesh to the Bay of Bengal. Many of Godavari's tributaries flow through Madhya Pradesh and Orissa making these states also co-riparians. The dispute arose when India's first Five Year Plan was being drawn up and the Planning Commission asked each of the states to name the projects they wanted included (Wood 2007: 66–69).

In July 1951, an inter-state conference on Krishna and Godavari rivers was held in New Delhi that included the then states of Bombay, Madras, Hyderabad, Madhya Pradesh and Mysore. But, the Government of Mysore subsequently did not ratify the agreement. However, in 1953 and 1956, major changes took place in the boundaries of the basin states during linguistic reorganization and it created upstream downstream conflicts. Therefore, various involved states applied for the setting-up of a tribunal between 1962 and 1969. Finally, in 1969 the central government set up two tribunals, one for each river, but consisting of the same personnel (ibid.).

The main issues to be decided by the tribunals were: (i) whether the agreement of 1951 was valid; (ii) whether Maharashtra could divert more Krishna water westward for its hydroelectric project at Koyna; (iii) how much water of both rivers should be allocated to each state for irrigation purposes; (iv) whether surplus water in the lower reaches of Godavari could be transferred by link canal to the lower reaches of Krishna, deprived of water by the Nagarjunasagar project (ibid.).

The awards of Krishna and Godavari tribunal in 1976 were as follows: (i) it declared the 1951 agreement invalid because Mysore (now Karnataka) had not ratified it; (ii) Maharashtra's claim for more water was rejected, and it declared that preference should be given to irrigation over hydroelectric generation; (iii) out of the available 2,060 thousand million cubic feet (tmcft) of river water Maharashtra would get 560 tmcft, Karnataka 700 tmcft and Andhra Pradesh 800 tmcft plus any excess, and projects begun before September 1960 should get preference. In the decades that have followed the Award, the basin states – particularly Maharashtra, Karnataka and Andhra Pradesh – have frequently been accusing each other over allocation of water and of violating the award (ibid.).

Although there have been a few significant disputes over Godavari water, recently the disputes over Krishna water has become a contentious issue. Andhra Pradesh complained that Maharashtra had violated the award by raising the height of the Ujjani dam and other diversions. Meanwhile, Andhra Pradesh announced 11 projects downstream from Alamatti, despite Karnataka's objections. Finally, a new Krishna Water Disputes Tribunal (KWDT) was set up in April 2004. At its first meeting Karnataka asked the new KWDT to restrain Andhra Pradesh from undertaking the projects (ibid.).

Another project that has aroused tension is the Telugu Ganga Canal, designed to transfer Krishna water to Chennai in Tamil Nadu. The agreement of 1983, which was presided over by Prime Minister Indira Gandhi, states that Andhra Pradesh, Karnataka and Maharashtra would each give 5 tmcft per year to Tamil Nadu's water-starved capital. However, it has not been fulfilled by the government of Andhra Pradesh yet. In September 2004, Chief Minister of Tamil Nadu wrote a letter to the Chief Minister of Andhra Pradesh, reminding him that Tamil Nadu had paid Andhra Pradesh ₹ 5.12 billion (US$ 115.2 million) but that 'even after two decades, the promised quantity of water is not reaching the Tamil Nadu border.' It clearly shows that pressure on water resources has increased among Godavari and Krishna co-riparian states (ibid.).

Cauvery River Dispute: The dispute over allocation of the waters of Cauvery River dates back more than a century. The river begins in western Karnataka and flows 381 km through that state, then forms the boundary of Karnataka and Tamil Nadu for

64 km, and finally descends to its huge delta in Tamil Nadu flowing 357 km through that state on its way to the Bay of Bengal. Kerala is a party to the dispute because it contributes to the catchment area of three of Cauvery's tributaries, and so is Pondicherry because sub-branches of the river irrigate its Karaikal area before reaching the sea (Benjamin 1971: 1794–1795; Wood 2007: 69–72; Chokkakula 2012: 29–34).

Cauvery is one of the most utilized rivers in the world. Barely 5 per cent of its water flows into the Bay of Bengal. The basin is known as 'the rice bowl of the south'. This is a classic case of upstream-downstream dispute mainly between Karnataka and Tamil Nadu. An agreement was made between the princely state of Mysore and Madras Presidency in 1892, which states that no irrigation reservoirs will be constructed without prior consent of the Madras government. However, both governments resented the agreement, and in 1924, a new final agreement was made for the extension of irrigation. On that basis, the Krishnarajasagar dam in Mysore and the Mettur reservoir in Madras were constructed. Mysore was allowed to extend its irrigated area but committed itself not to diminish the annual supply to Madras by more than 5 per cent. Madras committed itself to limit the area of irrigation under the Mettur reservoir. This agreement was to last for 50 years. However, after the end of the agreement in 1974, both side built structures to augment its share of water and accused one another of violating the agreement (ibid.).

In 1974, the Tamil Nadu government asked the central government to appoint a tribunal to settle the dispute. But the central government pursued the Government of Tamil Nadu to withdraw its request and suggested for negotiations. Two draft agreements were offered in the year 1974 and 1976, but were rejected and the water crisis prolonged in Tamil Nadu. Meanwhile, Karnataka, by adopting the delaying tactic of prolonging negotiations, was gaining time to enlarge or build new dams upstream. Finally in 1990, a tribunal was set up and in the year 1991, and an Interim Order (IO) was passed by the Cauvery Water Disputes Tribunal (CWDT), which states that Karnataka should ensure an annual release of 205 tmcft of

Cauvery water to Tamil Nadu (including 6 tcmft to Pondicherry). But, the Karnataka government argued that the IO was unfair and incapable of being implemented and proceeded to issue an ordinance against it to 'protect the interests of the farmers of the state'. The CWDT continued to work towards a final solution but the problem is still continuing (ibid.).

Ravi-Beas River Disputes: The Himalayan-fed Ravi and Beas rivers flow south-eastward across Punjab. Ravi arises in the Kulu Valley of Himachal Pradesh and it crosses the northern end of Punjab before forming India's border with Pakistan for some distance and then enters Pakistan near Lahore. Beas also begins in Himachal Pradesh and after entering Punjab becomes a tributary of Sutlej, a much larger river that originates in Tibet, crosses Himachal Pradesh and Punjab and also enters Pakistan (Wood 2007: 72). In 1955, at New Delhi, an inter-state water sharing agreement was reached to share the surplus waters of Ravi and Beas rivers. It involved the then States of Punjab, PEPSU (Patiala and East Punjab States Union), Jammu and Kashmir and Rajasthan. Rajasthan, though not a riparian state, was included as a beneficiary in then on-going Indus Treaty negotiations. In the estimated surplus waters of 15.85 million acre feet (maf), after deducting prior appropriation share of 3.13 maf, Punjab received 5.9 maf; Rajasthan 8.0 maf; PEPSU 1.3 M maf; and, Jammu and Kashmir 0.65 maf. However, the conflict over the water started to emerge soon after the states were re-organized. Punjab and PEPSU were merged as per the State Reorganization of 1956. Later in 1966, Punjab was divided into two states: Punjabi speaking Punjab and Hindi speaking Haryana. Therefore, the intervention of Central Government in 1976 and under Punjab Reorganization Act 1966, this share was divided with 3.5 maf to each State and the rest of 0.2 maf was given to Delhi. Punjab disagreed with the award and filed a suit in the Supreme Court challenging the decision and Haryana filed another suit to make Punjab implement the award. Hence, an agreement was made between Punjab and Haryana in 1981 with the intervention of the Central government. Under the agreement, Punjab received 4.22 maf and Haryana received 3.5 maf from a revised total available water of 17.17 maf. Later, a new dispute continued to simmer over implementation of the agreement and construction of Sutlej-Yamuna Link (SYL) canal, which would give Haryana access to its share. This figured in the Rajiv Gandhi-Longowal accord, and led to setting up the Ravi-Beas tribunal in 1985. The tribunal gave its award in 1987, and awarded 5.0 maf to Punjab and 3.83 to maf to Haryana. However, the dispute continued to recur. The prime reason of the dispute is that Punjab always opposed construction of SYL canal (Chokkakula 2012: 26–26; Wood 2007: 72–76).

Narmada River Dams

The construction of dams to harness the irrigation and hydroelectric potential has become controversial many a times in India. Among those, Narmada River represents India's most ambitious development project to date and also the most controversial. The Indian authorities planned numerous big dams for the Narmada River Valley as far back as 1946. However, there are few exceptions, few of these or other projects planned subsequently were completed by the early 1990s and other remained incomplete till today. Indeed, over 3,000 dam projects, including 165 big dams were finally approved for construction along the Narmada River's course and tributaries in 1979, constituting part of the largest river basin scheme formulated in India since Independence (Khagram 2004: 65–66).

There are number of references to irrigation development in the ancient writings about the Narmada. With the advent of British rule, irrigation in India gained more prominence, especially to repair and improve some of the most important existing irrigation work that had been indigenously built in the pre-colonial period. The First Famine Commission of 1880 further emphasized the need for direct involvement by colonial authorities in the development of new irrigation projects on the Indian peninsula (ibid.).

Yet, until the 1890s, the colonial government did not formulate comprehensive plans for any of the rivers on the subcontinent, partly as a result of the good agricultural harvest. Later on, the two great famines of 1887–88 and 1899–1900 subsequently compelled the British colonial authorities to appoint the

First Irrigation Commission to report on irrigation as a means for protecting against such disasters in India. The Commission conducted a thorough review of irrigation development and examined various proposals for new schemes to be built on India's rivers (ibid.: 66–68).

Even after the turn of the century and through the end of colonial period, no major irrigation works, such as canals or big dams were recommended for the Narmada River Valley, even though the British had built more than half of the world's big dam at the time. The Commission members argued that the deep black soil found in the basin would not likely stand up to the continuous irrigation made available by these types of large infrastructure projects. Thus, although a considerable amount of resources went into the promotion of smaller and private irrigation works, little activity in terms of large scale, state led development in the Narmada River Valley occurred until after World War II (ibid.).

Soon after Independence, under the leadership of Prime Minister Jawaharlal Nehru, the post-colonial Indian water and power bureaucracy forged ahead with the planning of big dams throughout India. The first proposal for the integrated development of the Narmada River Valley was made via the Indian bureaucracy in 1945–46 by Dr. A.N. Khosla, the Chairman of the federal Government of India's Central Waterways, Irrigation and Navigation Commission (CWINC). Correspondingly, the first formal investigations for comprehensive planning of the basin in the areas of flood control, irrigation, power and extension of navigation were initiated on the eve of Independence in 1946–1947 by the CWINC at the request of the Governments of the Central provinces and Berar and Bombay province. An ad hoc committee was made by the Central Ministry of Works, Mines & Power (CMWMP) to conduct preliminary investigation and to formulate developmental policy on the Narmada River Valley. It proposed the Bargi, Tawa and Punasa projects (the predecessor of the Narmada Sagar major dam project) in Madhya Pradesh, and Broach Barrage and Canal Project (predecessor of the Sardar Sarovar major dam project) in what was then Bombay

State. The Narmada basin appears to hold great potential for economic advancement of the country in general and the basin in particular (ibid.).

Narmada is the fifth largest river in the country with an annual flow of 36.5 maf almost equaling the combined inflows of Ravi, Beas and Sutlej. Originating in Madhya Pradesh, it has a catchment area of about lakh square kilometres and drains large parts of Madhya Pradesh, Maharashtra and Gujarat, before emptying into the Gulf of Cambay. Though one of the largest rivers, it is also the least developed (Menon 2003: 4095).

Complexities and Disputes

However, the project was delayed because of the involvement of multiple authorities and agencies within India's federal bureaucratic democracy in its formulation and implementation. The controversies go back to the early 1960s, when the initial problem lay in deciding how to allocate the costs and benefits of the project among the various claimant states. Thereafter, at the initiative of the then Union Minister of Irrigation and Power, an agreement known as Bhopal Agreement was arrived at in November 1963 by the Chief Minister of Gujarat and Madhya Pradesh. However, this agreement remained only on paper due to the subsequent rejection of the agreement by Madhya Pradesh. This necessitated the setting up a central committee headed by A.N. Khosla in 1964, but the Committee Report recommending a master plan for the basin also ended up in the government archives on account of disagreement among states (Menon 2003: 4095; Wood 1993: 968–969).

Most of the initiative for dam construction on the Narmada River has come from Gujarat and along with neighbouring Rajasthan, is one of the most drought prone states in India. Apart from Narmada, and Mahi and Tapi rivers, Gujarat has hardly any year-round rivers. Ironically, Narmada causes terrible floods in Gujarat during some monsoon seasons. The Gujaratis especially are India's most enterprising modernizers hoped that there will be agricultural and industrial growth by damming Narmada. They refer to Narmada, their only perennial river, as 'the lifeline of Gujarat'. On the other hand, people of Madhya Pradesh are eager to reap Narmada's benefits, but are preoccupied with other major river projects shared with other neighbouring states. Madhya Pradesh is less developed state compared to Gujarat to exploit Narmada and too lack resources to do it. The state of Maharashtra also has some interests to develop Narmada, but like Madhya Pradesh, it has other major rivers to exploit (Wood 1993: 971–972).

The never ending disputes among the states of Madhya Pradesh, Gujarat and Maharashtra led to the setting up of the Narmada Water Dispute Tribunal (NWDT) by the Government of India in 1969. Rajasthan was also made a party in the proceedings of the tribunal considering various issues. It issued its final decisions in August 1978 in a highly detailed five volume report (Menon 2003: 4095).

NWDT Award

The NWDT award spells out in great detail the apportionment of the costs and benefits of one of the world's most ambitious irrigation projects. According to planners, a network of 30 major dams (command area over 10,000 ha), 125 medium dams (400–10,000), and 3,000 dams (under 400 ha) would be built on the river and its 41 tributaries. Among those major dams, Sardar Sarovar dam in Gujarat is the major terminal dam on the main course of the river, the remaining 29 being in Madhya Pradesh. Sardar Sarovar dam, the world's longest cement lined canal, would stretch 300 miles northwest through central and northern Gujarat to Rajasthan. It was expected that 1.79 million hectares of land in Gujarat and 75,000 hectares in Rajasthan would receive from Sardar Sarovar, increasing more than six times at net value of their agricultural production. It would provide drinking water to 3,800 villages and 147 towns of Gujarat. It too would generate 1,450 MW of electricity (Wood 1993: 973–974).

However the effectiveness of Sardar Sarovar will depend on the upstream construction of the Narmada Sagar dam in Madhya Pradesh, which in turn will form part of the larger Indira Sagar project that includes three hydro dams: Narmada Sagar (1,000 MW), Omkareswar (520 MW) and Maheswar (320 MW). Narmada Sagar and Omkareswar will together irrigate 3,52,000 hectares of land in Madhya Pradesh, and Narmada Sagar's reservoir, which will store water for regulated release to Sardar Sarovar downstream, will form the largest body of water in India (ibid.: 973–976).

The most crucial decisions of NWDT involved (1) height of the terminal dam; (2) allocation of benefits; and (3) treatment of the displaced population. The height of the terminal dam (Sardar Sarovar) was set at 455 feet, with a takeoff point for the canal at 300 feet, which is less than what Gujarat wanted. However, it will create the second largest concrete gravity dam in the world, which will submerge a total of 34,996 hectares in Madhya Pradesh, Maharashtra and Gujarat, including 248 villages with a population of 66,593. Among these three states, Madhya Pradesh will bear the brunt of heavy loss of land, since 193 submerging villages are falling within its territory, populated of 45,000 people (ibid.).

As far as benefits are concerned, 28 maf of river flow, Madhya Pradesh was awarded 18.25 maf, Gujarat 9 maf, Rajasthan 5 maf and Maharashtra .25 maf. of the hydroelectric benefits, 57 per cent were allocated to Madhya Pradesh, 27 per cent to Maharashtra and 16 per cent to Gujarat. Considering the four-fifth of the Narmada flows within Madhya Pradesh, Gujarat will receive a disproportionate share of the water, but Madhya Pradesh has less need of water and is compensated with more electricity – more than it needs, in fact, enabling Madhya Pradesh to earn revenue by contributing to the all-India grid. These allocations will continue until the year 2025, and it may be renegotiated later on (ibid.).

Essential directions were made by NWDT for the displaced people about the issue of compensation, resettlement and rehabilitation. 'Land for land' was established as the basic principle of compensation to displaced families instead of cash compensations and they would receive land of their choice, which will be equivalent to their loss or a minimum 2 hectares in the irrigable command of the project. Moreover, every male of 18 years or older would be treated as separate family. It was also directed to give equal benefits to encroachers and the landless like those who have legal title over land. Approximately, two-third of the displaced people of Sardar Sarovar dam are tribal and therefore it was directed to rehabilitate them as whole in hamlets. Each family is to receive grant-in-aid money for resettlement and each resettlement village is to be provided with a primary school, a panchayat meeting place, a dispensary, a well link road, village pond and drinking water facilities, etc. (ibid.).

So far as the costs of projects are concerned, the NWDT allocated the costs to be borne by each of the states and set up the machinery to oversee the implementation of its decision. Since irrigation is a state subject in India, responsibility was given to Gujarat and Madhya Pradesh to establish the public corporations that would construct and manage the dam projects: the Sardar Sarovar Nigam Ltd., in Gujarat and Narmada Valley Development Authority in Madhya Pradesh. However, since Narmada is an inter-state river, the NWDT set up the Narmada Control Authority consisting of central and state engineers and bureaucrats to coordinate financing, construction and rehabilitation. Meanwhile, the centre controlled plans and expenditure through the Central Water and Power Commission, the Planning Commission, and the Ministry of Environment and Forests (ibid.).

The cost of building the Narmada projects, estimated at $15 billion (Sardar Sarovar alone is $3 billion), is to be borne by the state governments. However, in 1985, a loan of $450 million was provided to Sardar Sarovar by the World Bank. Further, World Bank loan of $350 million was granted for canal construction and $90 million for environmental protections (ibid.).

Although the NWDT took a long time to give its careful decisions in the year 1978, the controversies and conflicts didn't stop. It is true that the Tribunal's order determined many technical, financial and institutional parametres of the Narmada Projects, but the decisions about costs and benefits, resettlement and rehabilitation of the project affected people, environmental concerns/impacts were not properly addressed. The Tribunal also did not consider alternatives to the construction of big dams for development in the Narmada River Valley. On the other hand, a range of actors like World Bank, other foreign donors, domestic and transnational non-governmental organizations and people to be most negatively affected by the projects were not involved in the Tribunal's deliberations. In fact, the issue of resettlement and rehabilitation emerged as the centre stage of conflicts between the states (Khagram 2004: 76–83).

Addressing Displacement

The Narmada river basin is inhibited by 22 million people and among them 80 per cent of the people live in rural areas and depend primarily on agriculture. Within the basin, 245 villages, spread across three states, will be affected by the Sardar Sarovar dam – 193 in Madhya Pradesh, 33 in Maharashtra and 19 in Gujarat. The submergence area can be divided into two distinct socio-ecological settings. The villages in the lower hills of the river basin (upstream of the dam) are inhibited by tribal people. While the affected villages in Maharashtra and Gujarat have a near total tribal population, only about 65 villages in Madhya Pradesh fall in this category. These villages are marked by sparsely populated and subsistence oriented livelihoods, heavily dependent on rain-fed agriculture, livestock and forests. Almost every family owns some revenue land and 'encroached' forest land. Many of them have also migrated to Nimad plains, in Madhya Pradesh and Gujarat for wage employment (Dwivedi 1999: 47–48).

In the plains of Nimad, upstream, in Kukshi, Dharampuri, Barwani and Thikri sub-districts, affected area comprises approximately 120 villages. These villages are highly populated on account of its fertile land and the Nimad plain is known as the 'bread bowl' of central India. The region has undergone dramatic agrarian transformation, mainly because of 'pump revolution' bringing large area under irrigated agriculture. The major crops are bananas, sugarcane, papaya, cotton, chilly, soyabean and wheat. This area comprises big land owning community and middle peasantry. Rest of the population are the landless labourers (Scheduled Castes and Scheduled Tribes), constituting about 50 per cent of the population (ibid.).

The issue of human displacement and environmental consequences of major dams in India and the Sardar Sarovar project as particular attracted great attention. The Government of India contends that the large dam projects, like Sardar Sarovar, will provide drought-prone areas with irrigation and drinking water.

But, on the other hand, critics argue that by submerging vast amounts of land, the Sardar Sarovar displaces hundreds and thousands of indigenous people and causes large scale environmental destruction (Mehta 2005: 615–616).

It is well known that the displaced people, who are the citizens, rely on their own country's laws and policies with regard to resettlement and rehabilitation. However, there are no separate Indian laws pertaining specially to the state's legal responsibility to its displaced person. Thus, displaced must turn to the Fundamental Rights provisions of the Constitution and the writ jurisdiction. In fact, the Supreme Court of India has advanced human rights in India by implementing the principles of international declarations and treaties in support of the Constitution (ibid.: 627).

One of the most important provisions pertaining to human rights in the Constitution is Article 21, which provides the framework for securing the right to life. In addition, Article 39 directs the state to secure its citizens with the 'right to an adequate means of livelihoods'. In a related provision, Article 41 articulates that the state shall 'make effective provision for securing the right to work'. In an unprecedented decision, the Supreme Court of India expanded the socio-economic dimension of Article 21 in *Olga Tellis v. Bombay Municipal Corporation* and provided persons who are forcibly evicted with a more meaningful interpretation of the right to life. The court stated that any person deprived of his or her right to an adequate livelihood or right to work can challenge the deprivation as offending Article 21's right to life. In addition to interpreting Article 21 as recognizing the right to livelihood, the Supreme Court of India in *Mullin v. Union Territory of Delhi* elaborated on the right to adequate shelter as part of the all encompassing right to life (ibid.: 628–629).

The most important principle underlying the Land Acquisition Act (LAA) is the doctrine of eminent domain, by which the state can acquire land from private citizens for public purposes. The state's powerful right of eminent domain is nearly impossible to challenge legally, and thus the displaced are left with no mechanism by which to resist the state's acquisition of their land. The LAA specifies cash compensation at market value for the loss of individually-owned land and there is no provision of any kind of resettlement and rehabilitation of the displaced people. It has not specified rehabilitation is a prime consideration when acquiring land for a 'public purpose'. Finally, under the LAA, Supreme Court is the only appellate forum for the curtailment of the rights of the displaced people (ibid.: 629–630).

According to Supreme Court, the NWDT Award is the benchmark for the resettlement and rehabilitation, whose primary purpose was to establish conditions regarding the resettlement and rehabilitation of those displaced. However, the Award provides that relief is only available for 'Project Affected Persons' (PAPs), who are defined as families that lose their legally owned land to the reservoirs. The NWDT Award does not recognize families that lose their land to canals, suffer secondary displacement due to loss of their land, or those that are affected downstream of the Sardar Sarovar. It too does not provide for the tribal populations that cannot claim legal title to land for generations. In addition, the NWDT Award does not mention the status of those who are landless (ibid.: 630–632).

Unlike refugees and comparing to the Refugee Convention 1951, internal displacement persons do no benefit from an international legal and institutional framework. However, the United Nations Guiding Principles on 'internal displacement' was established in 1998. This is the first set of guidelines developed within the context of human rights law and international humanitarian law to address internal displacement and development induced displacement. The Guiding Principles are neither binding nor customary international law. However, it is useful because they establish that every person has a right to protection from displacement. Principle 7 of the Guideline stresses the need for concerned authorities to minimize the adverse effects of the displacement as well as to provide alternative accommodation of comparable value and use. Further, Principle 9 imposes particular responsibilities on states to prevent the displacement of tribal population, indigenous population and other such groups, given their unique reliance on their lands (ibid.: 621–624).

India has ratified the International Covenant on Economic Social and Cultural Rights (ICESCR) on 10 April 1979 and it holds India accountable. It states to recognize the right of all human beings to work, and the right to an adequate standard of living, including adequate food, clothing and housing. It says that any State action, which forcibly remove citizens from their homes are curtailment of these rights. Further, it declares that a state has prima facie failed to fulfil its obligation under ICESCR unless it can demonstrate that it utilized all available resources in an effort to secure these rights for its citizen. India too has ratified the 1957 International Labour Organization Convention 107 (ILO Convention 107) on 29 September 1958, was the first legal instrument to specifically provide such rights. Article 12(1) ILO Convention 107 says that indigenous people 'shall not be removed without their free consent from their habitual territories'. Further, Article 12(2) and 12(3) says to provide land of quality at least equal to that of the land previously occupied by the displaced, as well as full compensation for any resulting loss or injury (ibid.: 624–627).

Given the lack of comprehensive legal framework for resettlement and rehabilitation in the Indian context, the Guiding Principles, ICESCR, and ILO Convention 107 can provide a starting point for protecting displaced population in India. However, India's resettlement and rehabilitation of the people displaced by the Sardar Sarovar has been unsuccessful and inconsistent with all three international instruments. The Government of India has failed to provide sufficient compensation and proper resettlement, and has failed to protect the rights of tribal population (ibid.: 632–635).

It is the prime responsibility of the national government to protect the rights and provide assistance to the displaced. However, India lacks a comprehensive legal framework to define the rights of the displaced, and the solutions for rebuilding the communities and livelihoods of the people affected by the Sardar Sarovar project. The absence of state or national laws and policies to resettle its displaced repeatedly failed and the affected communities are forced to deal with sudden evictions, insufficient compensations, and the loss of their assets and livelihood. Despite the fact that the Sardar Sarovar has produced massive displacement and associated harms, the Government of India has yet to implement national and state legislation to protect those who encounter displacement against their will. However, in 1994, the Government of India has admitted that 10 million people displaced by dams and other developmental projects were still 'awaiting rehabilitation' (ibid.: 635–642).

Protests and Movements

Opposition was made by domestic groups in direct response to Tribunal's order in the state of Madhya Pradesh, particular case to Sardar Sarovar dam. The initial domestic mobilization in the state was the proposed height of 455 feet for the Sardar Sarovar dam specified by the Tribunal, which would cause the submergence of prime agricultural lands in the Nimar district of the state. The opposition was quickly expanded and intensified, because these lands were mostly owned by the middle-class farmers with substantial resources to deploy in social mobilization. In fact, cutting across the party lines, reactions were made against Tribunal's order across the state. A massive rally was organized by the 'Save Nimar' Committee in protest against the award (Khagram 2004: 83–84). This was the time; a new environmental awareness was spreading in India. Public concern was aroused by forests protection movements such as Chipko in Uttar Pradesh, Appiko in Karnataka, and anti-dam movements at Silent Valley in Kerala, Bedthi in Maharashtra, Bhopalpatnam and Ichhampali in Madhya Pradesh (Wood 1993: 976).

The opposition party (Congress party) in the state vehemently protested against the Tribunal's order and insisted the ruling government (Janata party) to stand against the order. Responding to the protests, the Chief Minister of Madhya Pradesh agreed to take action against Tribunal's order and filed a petition to lower the height of Sardar Sarovar dam to 436 feet in order to minimize the areas to be submerged in Nimar or risks its own survival in power. But the NWDT refused to reconsider this issue in 1979 and stated that re-opening of the hearings would have been against the procedures laid down by the Inter-State Water Dispute Act of 1956, which authorized only the clarification but not the re-adjudication of a Tribunal's decision once promulgated (Khagram 2004: 84–85).

However, the degree of mobilization of resistance was not so powerful enough to overcome the legal finality of the Tribunal's judgment and the 'Save Nimar Action Committee' too withdrew from the resistance believing that the Congress party leaders would continue to protest against Tribunal's order after being elected. However, it was a disappointment to say that; the Congress party after winning the election in Madhya Pradesh agreed to a political compromise and signed a tepid Memoranda of Understanding with the state of Gujarat (ruled by Congress party) to implement the decision of NWDT and to reduce the distress of displaced people as much as possible. The Government of Gujarat was very much enthusiastic to start the project and it was negotiating with foreign donors, primarily the World Bank, the United Nations and Japan for assistance, even before the Tribunal's final order in 1978. In November 1979, prior to the Tribunal's final award declaration, the Government of Gujarat played host to the World Bank, which had been watching the Tribunal's hearings with great interests (ibid.). Credit and loan agreement was signed in 1985 between the World Bank and Gujarat, Madhya Pradesh and Maharashtra to start the project (Wood 1993: 976).

However, the Sardar Sarovar project started only after it got environmental clearance from the Ministry of Environment and Forests, Government of India. The Department of Environment was created in the year 1980 and in the same year, the Parliament passed the Forest Conservation Act giving the Department (later Ministry) of Environment and Forests (MEF) the authority to demand an environment assessment of development projects before central clearance. In the case of Narmada projects, Gujarat and Madhya Pradesh were required to (1) prepare alternative afforestation sites to compensate the submerged forests land; (2) improve the rehabilitation package for displaced people; (3) create sanctuaries for wildlife; and (4) prepare both command and catchment areas for the dams' negative environmental effects. By 1983, when the Sardar Sarovar and Narmada Sagar projects came up for clearance,

clearance was refused by the MEF, because it did not meet MEF guidelines. Finally, after several costly drought years in Gujarat and increased political pressure in that state, the Sardar Sarovar and the Narmada Sagar projects were given conditional clearance by the Government of India in June 1987 and construction work began immediately on Sardar Sarovar (ibid.).

After getting the clearance, the Bank's mission subsequently made three recommendations: (1) establishment of a high level Narmada Planning Group in Gujarat to coordinate the project; (2) the retention of domestic consultations to conduct various studies; (3) the acquisition of foreign experts to investigate specific technical issues for which domestic consultations were not available. The World Bank secured $10 million from United Nations Development Programme to assist the Government of Gujarat in implementing these recommendations (Khagram 2004: 86).

The issue of resettlement and rehabilitation, particularly of indigenous or tribal peoples, and environmental concerns began to emerge as critical issues on the international and Indian developmental agendas. The Tribunal's ruling on resettlement and rehabilitation had mandated land and cash compensation for each family, continued support through grants and financial assistance, and relocation of communities as units to new villages. However, implementing the project of resettlement and rehabilitation as per Tribunal's order was quite tough for the governments concerned. First, the bureaucratic machinery failed to implement the Tribunal's order, since the execution of the project ultimately depends on the cooperation of state and federal governments. Second, the question arise what to do with encroachers on State-owned forest and wastelands (of whom many were tribal peoples who did not have land titles) was not explicitly addressed. Finally, the award specified the Resettlement and Rehabilitation package for the displaced of Maharashtra and Madhya Pradesh, and for the Government of Gujarat's assistance to those States, but made no mention of the package for the displaced of Gujarat itself (ibid.: 87).

Therefore, revival of protests is obvious. Several years earlier, there were long struggles to ensure the displaced population for proper compensation, resettlement and rehabilitation. One of the earliest activist groups in Gujarat was the Chhatra Yuva Sangharsh Vahini (Student-youth struggle) formed to fight for the people from 19 Gujarat villages to be submerged by the Sardar Sarovar Reservoir. It forced the Government of Gujarat to improve the rehabilitation package. Its members too assisted directly in the resettlement process, attending to the welfare of families and insisting on full implementation of agreed-upon terms. Aware of Gujarat water needs and of the strength of pro-dam sentiment in the state, the *Vahini* and other volunteer agencies there maintained a tough 'watchdog' stance while operating with government rehabilitation efforts (Wood 1993: 977–979).

On the other hand, there was an anti-dam sentiment in Maharashtra and Madhya Pradesh and the protest groups were more radical. They were created and led by human rights and environmental activists who reject the project for its disastrous consequences for the Narmada Valley and its residents. The groups included the Narmada Ghati Navanirman Samiti (Narmada Valley new awakening committee) in Madhya Pradesh and the Narmada Ghati Dharangrastha Samiti (Narmada Valley committee for the dam affected) in Maharashtra. Later, these merged to form the Narmada Bachao Andolan (Save the Narmada movement). Their activities included consciousness-raising among tribals, *rasta roko* (road block) and other protests efforts. Many a times, they staged demonstration against the World Bank on numerous occasions. In the year 1989, the protest got more momentum when the famous leprosy worker and social activist, Baba Amte, joined their ranks and led an anti-dam rally of 60,000 at Harsud in Madhya Pradesh, a town of 20,000 slated to be flooded (ibid.).

Within a short period of time, protest against Sardar Sarovar project was too supported by various national and international organizations. A transnational coalition was set up among various national and international non-governmental organization and human rights organizations. Such are Oxfam, Survival International, The Ecologists, Arch Vahini, Lok Adhikar Sangh in Gujarat, Kalpavriksh in New Delhi, and others. From

1984 on, the federal and state governments in India as well as the World Bank began to face increased pressure from above by foreign and international non-governmental organizations, as well as from below by grassroot groups and domestic non-governmental organizations and the increasing transnational links between the two (Khagram 2004: 90–91).

The World Bank Independent Review

These transnational lobbying and advocacy efforts gradually began to have an effect on the World Bank. During 1991, under pressure from governments of several member countries, the World Bank decided to commission an independent review of the environmental and resettlement aspects of the Narmada Development Project. The commission submitted its report 'Sardar Sarovar: Report of the Independent Review' in less than 10 months. The main findings of the review can be summarized as follows. First, with respect to resettlement and rehabilitation of the displaced people, it states that neither the Indian governments involved nor the World Bank carried out adequate assessments of the human impact of Sardar Sarovar. There was also no consultation with the people to be affected by the project. Second, the review states that there was no recognition of the special needs of the tribal populations involved. Third, it reports that the compensation packages for Madhya Pradesh, Maharashtra and some Gujarat 'oustees' did not meet World Bank standards and it expresses its doubt as to whether Gujarat could absorb the large number of people to be resettled from Madhya Pradesh and Maharashtra. The report too received attention on two new categories of people. Such are the affected people in the downstream and their major concerns of loss of their source of livelihood and people affected by canal construction and dam's infrastructure. The review argues that they should be compensated according to the same terms as those whose lands are submerged (Wood 1993: 979–982).

Concerns on environmental effects, the report states that the environmental impact assessment studies were inadequate and there was no proper plan to mitigate environmental damage. Further, the people in the valley and the canal vicinity were not consulted while preparing such a plan. At the same time, it strongly criticized the existing schemes for compensatory afforestation, prevention of water logging and salinisation in the command area, distribution of drinking water and malaria control etc. The review too questioned the hydrological assumptions on which construction of the dam and canal was based, arguing that they would not perform as planned since river flow was less what the NWDT calculated it to be (ibid.).

Indicating all these shortcomings, the review called for the World Bank to 'step back' from the Narmada projects and stated that proper human and environmental impact studies need to be completed and World Bank policies on resettlement and rehabilitation must be reviewed. It reported that the World Bank should suspend its funding of the projects until the governments of India and the involved states complied with expected standards of oustees rehabilitation and environmental protection, which had set themselves in various undertakings in Narmada projects, including credit and loan agreement with the World Bank (ibid.).

However, the report was quickly castigated as 'lacking impartiality', and many condemned it as a part of conspiracy to block India's progress. The Chief Minister of Gujarat rejected the review's findings, complaining that they reflected only the views of the dam's critics. The Minister of Irrigation, Madhya Pradesh, called the report 'willfully motivated' by 'vested interests' lobbying against the project at home and in international fora (ibid.).

Eventually in 1993, the World Bank withdrew itself from the project. However, Indian government opted to continue without Bank funding despite citizen protests, international condemnation, a sizeable decrease in financial support, and substantial concern over the project's impact on the environment and human displacement. Gujarat official stated that World Bank aid represented only 10 per cent of the total cost of Sardar Sarovar and even if the Bank withdrew itself, construction would continue even if delayed (Mehta 2005: 617–620).

While the Bank's withdrawal constituted a significant milestone in the history of the project, it also diminished pressure on the project-implementing authority agency to adhere to international and environmental requirements. The Indian government's decision to pursue the project without the Bank's involvement forced Sardar Sarovar Project resistance movements to mobilize its forces at the domestic level. In 1994, the Narmada Bachao Andolan filed a public interest petition with the Supreme Court of India. The Government of India halted work on the dam in 1995 due to a Supreme Court ruling recognizing that the rehabilitation of displaced people was inadequate (ibid.).

The Supreme Court Decision

The Supreme Court (SC) made its judgement in the case of *Narmada Bachao Andolan v. Union of India and Others* on 18 October 2000. The six years (1994–2000) between the filing of the NBA's writ petition and the SC judgement were among the most politically chaotic and uncertain in India since Independence. Three general elections in four years reflected an uncertain and fragmented electorate, and continual coalition formation, break-up and reformation resulted in weak governance. One of the consequences of this political disarray was a new assertiveness on the part of India's judiciary, particularly at its highest level, the Supreme Court of India. The SC had begun to expand its role in the late 1970s by encouraging public interest litigation (PIL), that is, class action suits aimed at promoting social and economic justice. It was widely admired for taking up issues that politicians and the political system were incapable of resolving. However, judicial activism also had its critics and opponents. The judges said that they had reached too far beyond their normal sphere of competence and become overly ambitious in trying to effect social change. In this process, it had stimulated an immense increase in India's already staggering litigation backlog. By 2000, it was a worry for judges that the court could lose their reputation for impartiality and integrity, without which their effectiveness would be seriously impaired (Wood 2007: 171–172).

Nonetheless, the NBA had high expectations, not only that the SC would find in its favour regarding the injustices on the Sardar Sarovar oustees and the environment, but also that the

Court would stop the building of the dam altogether. When the SC issued its stay order on Sardar Sarovar construction in January 1995, the NBA appeared to be at the height of its success. The struggle of NBA had spread worldwide through print and visual media and, outside Gujarat. It appeared that the NBA was winning the battle of Indian public opinion regarding big dams and water resource development (ibid.: 172–173).

By the end of the 1990s, the NBA's struggle was boosted by other like-minded organizations. Internationally, 'sustainable development' had become the watchword of all aid donor organizations, conferences, seminars, books and articles dealing with the Third World. In April 1997, the World Conservation Union (IUCN) held an International Workshop on Large Dams at Gland, Switzerland, attended by stakeholders from 18 countries representing all shades of pro- and anti-dam opinion. The Workshop sought a consensus on 'whether dams have a major role to play in land, water and energy development and how particular dams should be selected, constructed and operated.' The report of The World Commission on Dams (WCD) was eventually published as *Dams and Development: A New Framework for Decision-Making*, and widely circulated. Although Narmada dams were not featured in the six major case studies of dam projects in different parts of the globe, India and China were asked to prepare country reports that provided comprehensive analyses of their entire dam building experience. The Indian report (*Large Dams: India's Experience: A Report for the World Commission on Dams*) noted that large dams in India have made important contributions to the development of irrigated agriculture, improving food production as well as contributing hydro-electric power and enhanced domestic and industrial water supply. On the other hand, it has stated the socio-economic, environmental and financial costs of the Indian dam projects. It concluded that many Indian dam projects undertaken since the 1980s were, and remain, unviable. Despite the pros and cons of dam projects, both the Indian and WCD reports that ways can be found to improve dam-building so that the injustices and harm they have inflicted can be minimized, if not eliminated altogether (ibid.: 173–174).

On the political front, the hopes of the people of Gujarat for an end to the Sardar Sarovar stalemate rose when the BJP formed a coalition government in New Delhi in 1999. Throughout the election campaign, BJP declared that if the BJP were elected, the Narmada dams would become a 'National Project', implying that all obstacles would be removed to complete the Sardar Sarovar dam. Meanwhile, the BJP government in Gujarat was renewing its efforts to deal with the rehabilitation and resettlement of oustees. The new minister in charge of Narmada Development, Jainarayan Vyas, announced on 16 March 1998 that the Gujarat government would allow 8,000 project-affected families in Madhya Pradesh to decide where they would like to relocate. The announcement showed that the Gujarat government was doing its utmost to impress the SC with its reasonableness in dealing with rehabilitation and resettlement (ibid.: 175–176).

The most convincing evidence of this reasonableness came in February 1999, when the Government of Gujarat created the Grievances Redressal Authority (GRA) for Sardar Sarovar Project Affected Persons. This body was to provide a means whereby oustees resettled in Gujarat could get redressal for their grievances until the entire resettlement process was completed P.D. Desai, retired Chief Justice of the Bombay, Calcutta and Himachal Pradesh High Courts, was appointed the chairman of the GRA. Desai quickly established a reputation for integrity and fairness. According to GRA records, between April 1999 and December 2000, the GRA dealt with 14,158 grievances; out of these, 10,725 were decided in favour of the oustee complainants. The work done by GRA was well appreciated by the Government of Gujarat. Perhaps the greatest compliment paid to the GRA came, when the governments of Maharashtra and Madhya Pradesh ordered that GRAs be set up in their states as well (ibid.).

It has been mentioned earlier that the SC made its judgment on 18 October 2000. Justice Kirpal rejected the petitioners' claim and it stated that the displacement caused by Sardar Sarovar does not violate the fundamental rights of the oustees. Rather, he maintained that the project is highly needed to quench the thirst of water for burgeoning population. It is highly needed for food security, domestic and industrial water supply, hydro-electric power and flood control. He said, once the oustees were rehabilitated at a new location, they would be in a better position than what they were in earlier. They will avail better amenities and facilities of life as compared to their tribal hamlets. Their gradual assimilation in the mainstream of the society will lead to betterment and progress (ibid.: 180).

With regard to the environmental issue, the NBA objected the environmental clearance of Sardar Sarovar and stated that, giving clearance in 1987 was political and, thus, it violates the rights of the affected people under Article 21 of the Constitution. It argued that the studies that had been carried out were of poor quality, based on improper data. However, Justice Kripal rejected the claim and stated that the collection of impact data was adequate. He said that care for environment is an ongoing process and the system in place would ensure that ameliorative will be taken to counter such adverse effects.

The main contentions of the NBA regarding the oustees were, first, that the number of affected families by dam construction has increased drastically and they need to be additionally compensated. Second, there was no master plan for resettlement and rehabilitation. Finally, the existing R&R sub-group of the Narmada Control Authority (NCA) as governmental agency did not have the requisite independence to monitor the R&R provisions, which had made in NWDT Award (ibid.: 180–182).

Justice Kripal admitted that the number of Partially Affected Families (PAFs) had risen from 7,000, as stated in the NWDT Award, to 40,227 as of 1990. The reason for the rise in numbers was principally because of the liberalization of the eligibility rules for R&R benefits. Justice Kirpal rejected the idea that PAF status should be extended further. With regards to the differential in the R&R packages offered by Gujarat, Maharashtra and Madhya Pradesh, Justice Kirpal argued that there is no need to force the three states to offer identical packages; oustees were free to choose the option that worked best for them. He cited important inter-state meetings in 1996 and 1999 where commitments had been made to review the implementation of R&R measures. Justice Kirpal reviewed and approved all the measures designed to protect the welfare of the oustees; to preserve their social and cultural environment, provide civic amenities, improve agricultural production and extend health care facilities, etc. Finally, Justice Kirpal said that the R&R subgroup and the rehabilitation committee of the NCA constituted a well established mechanism for coordination and monitoring of R&R programmes. He said that, it would closely monitor R&R progress with regard to land acquisition, rehabilitation of PAFs and their implementation. He saved his strongest commendation for the GRA, which he described as 'an effective monitoring and implementing agency with regard to relief and rehabilitation of the PAFs in Gujarat' (ibid.: 182–184).

The SC judgment on Sardar Sarovar was warmly greeted by the Government of Gujarat, while the NBA activist Medha Patkar said, the re-launch of the construction of Sardar Sarovar was a 'manifestation of the arrogance of state power which would be resisted by the people of the Narmada valley'. It was also criticized by two former administrators (Ramaswamy Iyer and L.C. Jain) who had been members of the Five Member Group (FMG). Iyer argued that it was wrong to trust the NCA and its sub-groups to monitor the fulfillment of R&R. He objected to the judges' advocacy of dams in general and stated that the justices were unfair to the NBA. Iyer found it 'unfortunate and disquieting' that all the good work of PIL over the previous two decades 'could be nullified at one stroke by this judgment.' This judgment is a severe blow to people's movements. L.C. Jain said the judgment is pro-development and pro-large dams and viewed that the displacement of *adivasis* was a 'desirable goal in itself'. What he found most disturbing in the judgment was the knowledge-base of the judges who had discounted all the negative evidence against large dams. The majority judgment seemed blind to the controversies, such as the utilizable quantum of water in the river or the height of the dam and etc. He said that the NBA is treated unfairly by the Court (ibid.: 188–190).

The case of *Narmada Bachao Andolan v. Union of India and Others* seems likely to become a landmark in the history of river water dispute resolution in India. Although the SC may not have delivered the kind of justice sought by human rights and environmental protection activists, it did not just reaffirm the status quo. The judges, in effect, said that the existing institutions for water dispute resolution – supplemented by the new GRAs – were adequate to deal with conflicts over the development of India's river water resources (ibid.: 192).

5

Hirakud Dam and Development

Turbulent Rivers and Floods in Orissa

Economically, Orissa is one of the most backward states in India and it has been proved repeatedly in terms of its several socio-economic indicators over the last few years. However, apart from socio-economic factors, natural factors are also responsible for its backwardness (Baboo 1991: 2373). Generally, rivers provide the most efficient system of drainage, but in event of occurrence of floods they upset peoples' economy and disturb their lives as well and Orissa is not exception to it (Agarwal 1967: 1–2).

The state of Orissa is traversed by five rivers, such as the Mahanadi, the Brahmani, the Baitarani, the Burabalong and the Subarnarekha. The first three are the principal rivers of Orissa, originating from the hilly terrain of the Central Provinces and run parallel to each other and join the sea directly. Among them, Mahanadi is the largest river which rises near Sihawa in the extreme south-east of Raipur District in Central provinces (now Chhattisgarh plains). After a course of about 470 miles, the Mahanadi pours down upon the Orissa delta from between two hills at Naraj about seven miles west of the city of Cuttack. It traverses Cuttack district from west to east and after fragmenting into various distributaries at a distance of about 50 miles from the sea, falls into the Bay of Bengal through several channels near False Point. The Brahmani is formed by the junction of the South Koel and Sankh rivers in Gangpur and following a very winding easterly course it is joined by Baitarani and reaches the Bay of Bengal by the Dhamra estuary. The Baitarani rises in the north–west of Koenjhar, and flows first to the south–west and then takes an east turn. It is joined by the Brahmani, following which the united stream now called Dhamra falls into the Bay of Bengal (Mahalanobis 1940: 1–2).

The primary cause of floods in Orissa is the huge volume of water poured into the delta through the Mahanadi, the Brahmani, and the Baitarani during periods of excessive rainfall in their respective catchments areas. Each river enters Orissa in one single stream and almost immediately divides into a great number of branches owing to deltaic action. The velocity is suddenly reduced on reaching the plains and the silt is deposited on the river beds. In this way the bed of the river rises and gradually a shallow ridge is built up, on the summit of which the river flows. Fresh ridges are formed by the deposit of silt along the channel of new branches, and the whole process is repeated with the formation of more new branches. During heavy floods, water is spilled over either of the banks and silt is deposited over the surrounding country which slowly raises the general level of the deltaic land. Such deltaic actions in all the three rivers have further led to creation of 15 main branches along with innumerable minor branches about half way between gorges and sea (ibid.: 3).

Although the total area of the delta is about 8,000 square miles, the central area of about 3,000 square miles is the most important. The greater part of the central area that is 2,300 square miles in extent consists of the Mahanadi delta, while the north central area of about 800 square miles is watered by the distributaries of the rivers Brahmani and Baitarani. The central area is the most fertile and densely populated tract in Orissa and has a high intensity of floods which is largely due to the land being extremely flat. The deltaic channels, in spite of their great number, are unable to carry away even half the water brought down by the rivers from the western mountains during high floods. Owing to the insufficient carrying capacity of the drainage channels, the flood water cannot quickly escape into the sea (ibid.: 3–4).

The situation is further aggravated by obstruction to a free flow cause by canals and embankments constructed by men for irrigation and navigation. Another factor of importance is the action of tides, which at times holds up the water and appreciably prolongs the duration of floods. Tidal action extends to a considerable distance inland and causes marked variation in the level of all the rivers. During the monsoon season, at the mouth of the river the water level in the Bay of Bengal rises to about 3 feet or more during storms. Owing to the very small width of the Orissa delta (from 30 to 50 miles), the tidal action offers considerable obstruction to the flood discharge by appreciably reducing the slope of the water flowing down the rivers (ibid.: 4).

Mahanadi Valley Development: The Unified Basin-Wide Plan

The occurrences of frequent floods in Orissa pushed the government to formulate a scheme in the year 1947, known as Mahanadi Multi-Purpose River Valley Development, a 'Unified Basin-Wide Plan', to put a stop to colossal waste of the natural resources of Orissa. The scheme was designed to control, conserve and utilize the water of Mahanadi River primarily for flood control. Its auxiliary functions included irrigation, hydropower generation and navigation. The unified basin-wide plan for the development of Mahanadi valley comprised construction of a series of dams on the main river. According to the plan, three units would be constructed i.e. the Hirakud dam project, the Tikarpada dam project and the Naraj dam project. These units would be individually capable of independent development, irrespective of whether other two dams were constructed or not, but would form an integral part of unified basin wide plan. The best result of flood management would obviously be obtained if and when all the three projects were completed. However, the overall development of this plan would take time to complete; and therefore it was proposed that the Hirakud Dam Project could be initially started (Government of India 1947: 11–15; Mahalik 2005: 75–76).

Out of the three units of the basin-wide plan, Hirakud dam project lies in the upper most part of the Mahanadi River and its foundation stone was laid by Sir Howthrone Lewis, the then Governor of Orissa on 15 March 1946. The project report was submitted to the Government in June 1947. Its construction was started in 1948 and was completed in 1956. Later, it was formally inaugurated by Prime Minister Jawaharlal Nehru on 13 January 1957 (Mahalik 2005: 76–78). It is located across the Mahanadi below the confluence of 'Ib' River, about nine miles upstream of the town of Sambalpur. It is two miles wide and is flanked by two hills. One arm of the dam joins the hill named

Chandli Dongri which rises steeply from the river bed on the right side. The left arm of the dam joins the hill known as Lam Dungri which is situated at a small distance from the river bed (Agarwal 1967: 24–33).

Flood Management

It has been discussed earlier that the flat nature of the delta and the shallow depth of the river channels of Mahanadi made the waters overflow and flooded the banks during the rainy season (Agarwal 1967: 40). The Hirakud dam project for the first time offered an effective solution to the problem of floods in the delta (Mahalik 2005: 77). The dam was expected to benefit the delta in two ways; first, it would save damage to lands and property; second, it would make land fit for irrigation and cultivation (Government of India 1947: 50). However, it did not live up to these expectations. Although a large dam was constructed on the River Mahanadi to check floods, the situation did not change and remained the same. After the construction of the Hirakud dam, Orissa has been affected by floods 26 times within 55 years. There have been 20 high floods and 8 major floods in the period 1958 to 2011. It has been reported that, floods have taken 851 human lives and 74,621 livestock during the period 1960–2006. It has also severely affected cultivated lands ranging from 1 lakh to 16 lakhs hectare of land in each flood and brought the loss of ₹ 50.5794 billion rupees during the same period (Choudhury, Sandbhor and Satapathy 2012: 10–20).

Government record shows that river Mahanadi is causing high floods almost in every three years. The coastal districts have been experiencing similar situations. This flood situation could largely be attributed to the nature of the rainfall. Both, the Hirakud dam and the downstream of the reservoir get flooded as heavy rains inundate the area. Such occurrences have been experienced in 1958, 1959, 1961, 1992 and 1994. Downstream of the reservoir or the coastal Orissa gets flooded when heavy rains occur only in the downstream. All the downstream tributaries of Mahanadi supply huge volumes of water to Mahanadi and it gets flooded. Such occurrences were observed in 1969, 1982, 1991, 1992, 2001, 2003 and 2006. On the other hand, Hirakud dam gets flooded only when it rains heavily in the catchments. All the upstream tributaries of Mahanadi supply enormous volume of water to Mahanadi and Hirakud dam due to which it gets flooded. Such instances were documented in 1959 and 1980 (Table 1). The study of D'Souza (2006) too claims that the Hirakud dam on the river Mahanadi has altered the hydraulic character of the coastal plains in Orissa. Instead of managing floods in the coastal area, the hydraulic interventions (Hirakud dam) transformed the Orissa delta from being a flood dependent agrarian regime into flood-vulnerable landscape (D'Souza 2006: 216).

Table 1

Post-Construction High Floods above 10.00 Lakh Cusecs from 1958–2006

S. No.	*Year*	*Month of Occurrences*	*Duration of Flood Period*	*Remarks (Reason of Floods)*
1	1958	July	10/07 to22/07	B
2	1959	July	04/07 to13/07	H
3		Sept	10/09 to 18/09	B
4	1961	July	16/07 to 26/07	B
5		September	01/09 to 11/09	B
6		September	14/09 to 17/09	B
7	1969	August	01/08 to 06/08	D
8	1980	September	19/09 to 30/09	H
9	1982	August–September	30/08 to 04/09	D
10	1991	July–August	08/07 to 02/08	D
11		August	12/08 to 21/08	D
12	1992	July	08/07 to 19/07	B
13		August	16/08 to 30/08	D
14	1994	August–September	27/08 to10/09	B
15	2001	July	04/07 to 31/07	D
16	2003	August–September	28/08 to 15/09	D
17	2006	August–September	30/08 to 04/09	D

N.B: B – Both For Hirakud Dam and Downstream
H – For Hirakud Dam Only
D – For Downstream Only

Source: Government of Orissa (2006), *Flood Report of Hirakud Reservoir for 2006*, Department of Water Resources, Superintendent Engineer Hirakud Dam Circle, Burla, p.132.

As mentioned above, the unified basin-wide plan for the development of Mahanadi Valley had envisioned construction of three dams on the main river and its major tributaries to check floods. However, only the Hirakud dam project has been completed till now and the proposals for other two dams have been sidelined. Not surprisingly therefore, Hirakud dam alone is unable to control the floods. Clearly the absence of other two projects – one at Tikarpara and another at Naraj – is one of the major reasons of failure of flood control, which was one of the major objectives of Mahanadi Valley Development plan (Ramdas 2001).

Additionally, high rate of silt deposits in the Hirakud reservoir is also a huge contributing factor which is gradually reducing its flood control capacity. The high rate of silt deposit reduces the reservoir storage capacity and thus increases the level of floods. Silt observation of the Mahanadi at Hirakud was started on 1 October 1946. According to initial estimates, the annual silt deposit was expected to be between 8,000 to 12,000 acre ft or 0.008 to 0.012 Macft per year, thus giving the reservoir a life of 100 to 150 years. On the conservative side, it was taken to be 100 years (Government of India 1947). However, the post-construction era has shown that, the rate of silt deposit has been unexpectedly high making flood control its obvious casualty. The annual silt deposition in Hirakud dam is 0.023 against the annual rate of 0.012 Macft per year. It means that it has been depositing an extra 0.011 Macft per year, which is 91.6 per cent more than the initial prediction. The dead storage capacity has been reduced from 1.880 to 0.870 Macft within 43 years (1957–2000). Alternatively, it means that if the rate of deposition of silt continues at the same rate, the reservoir has only 28 more years of life from the year 2010 (Table 2).

This high rate of silt deposition is due to the large scale industrialization and deforestation in the catchment area of Hirakud dam. A numbers of industries have been established in the catchment area especially lured by the easy and cheap power and water supply by Hirakud dam. After the completion of Hirakud dam in 1958, the most devastating floods occurred in the years 1961, 1982, 1994, 2001, 2003 and 2006. The first shocking flood of 1961 prompted the government of Orissa to

Table 2

Deposit of Silt in Hirakud Reservoir

Survey	*Dead Storage Capacity (Macft)*	*Reduction of Dead Storage From the Original Project of 1957 (Macft)*
Original 1957	1.880	
Third Cycle Hydrographic Survey 1986	1.223	0.657
Recommendation of the Rule Curve Revision Committee 1988	1.472	0.408
Remote Sensing Survey By Central Water Commission 1995	0.980	0.900
Sedimentation Survey 2000	0.870	1.010

Source: Government of Orissa (2006), Department of Water Resources, Superintendent Engineer Hirakud Dam Circle, Burla.

immediately start the construction of second phase of the unified basin-wide plan of Mahanadi Valley Development. Prime Minister Jawaharlal Nehru got the Tikarpara project cleared and laid the foundation in early 1964. But the project failed to take off, because of stiff resistance from the people of western Orissa who feared that it would submerge vast areas including the newly irrigated areas of the Hirakud dam and the historical towns of Boudh and Sonepur. Over two decades later, in 1982, the Mahanadi got flooded again. The government of Orissa again revived the question of a second dam on the Mahanadi. The government argued that a dam at Tikarpada, as proposed earlier, would lead to the submergence of historical towns, and wanted it to be built at Manibhadra. A report was also prepared, but the project is yet to take off (Ramdas 2001).

Hydropower

Most of the population of Orissa is occupied in agriculture and the standard of living is very low due to its exclusive rural economy. Although Orissa is deposited of varieties of minerals (iron ore, manganese ore, bauxite, limestone, dolomite, chromites, coal and graphite) and in large quantity too, it was lacking both cottage and large industries to exploit those. Orissa had the potential to become a great centre of industry, which would accelerate the standard of living of people, if its resources were utilized with adequate plan. Therefore, electricity is highly needed and plan was made to harness it from Hirakud dam. Many proposals were made in the feasibility report of Hirakud dam to open up various industries at different places of Orissa. Such are aluminum factory at Hirakud, ferro manganese plant at Joda and Hindustan Steel Ltd. at Rourkela.

Paper mill at Chowdwar and Brajaraj Nagar, cement factory at Rajagangpur, textile mills, sugar mills at Bargarh, tube mill at Chowdwar and development of mines and collieries. There will be subsidiary industries around all these new industries and be remodeled with the help of cheap electricity. Demand for electricity both in rural and urban area would also be fulfilled to refashion the way of life. The people will get more employment in the newly established industries and the per capita income would go up. There will be considerable stimulus to trade and transport services and economic and social activities. In short, the generation of hydropower from Hirakud dam would develop the all round prosperity in Orissa (Agarwal 1967: 63–66).

Orissa is the most underdeveloped state in the Indian Union with the lowest per capita consumption of 0.29 kWh of electricity as against 9.2 kWh for the whole country (Agarwal 1967: 63). In order to bridge the gap, it was proposed to install a capacity of 3,50,000 kW at the Hirakud dam to carry a peak load of 3,00,000 kW. The power generation will be done in two parts: one at the main dam at Hirakud and second at the subsidiary dam, forming the balancing reservoir at the end of the tail race from the upper power house, to utilize the steep slope in the river below Hirakud. For the total installed capacity of 3,50,000 kW, it was proposed to install 6 units of 25,000 kW each at the upper power station and 8 units of 25,000 kW each at the lower power station. Provision had also been made to raise the installed capacity subsequently to 4,75,000 kW, if the demand increases. For the ultimate installed capacity of 4,75,000 kW, there will be 8 units in the upper and 11 units in the lower power stations. But the financial forecasts deal during that period was meant to the installed capacity of 3,50,000 kW only and would generate 1, 524 million units of hydro-power (Government of India 1947: 59–61).

As per the original project report (1947) of the Hirakud dam project, the power generating installed capacity was proposed of 3,50,000 kW (350 MW), comprised two power stations; one at Burla (upper power station) and another at Chiplima (lower power station) (Government of India 1947: 59). But at present, the total installed capacity of Hirakud project has only reached to 307.50 MW. These were installed under three phases (Murty and Mishra 1994: 132; Orissa Hydropower Corporation Ltd. 2007). However, even after 50 years completion of Hirakud dam and its installation of three stages of hydropower the target have not been fulfilled. It lacks 42.5 MW of installation capacity to reach its target. Hirakud dam fulfilled 35.14 per cent of installed capacity in the year 1956, 77.14 per cent in its second installation in the year of 1964 and 87.86 per cent in its third installation in the year of 1990. It still lacks 12.14 per cent of installation capacity to reach its target (Orissa Hydropower Corporation Ltd. 2007).

The first hydropower station of Hirakud dam was commissioned in 18 December 1956 and as per the proposal of the feasibility report, a large number of new industries were established and came into operation across Orissa. Such are cement factory at Rajgangpur, steel plant at Rourkela, Indian Alumina Co. at Hirakud, ferro manganese plant at Joda, Titagar Paper Mills at Chowdar, refractories at Belpahar, Orient Paper Mills at Brajarajnagar, etc. The hydropower system of Hirakud and its operation became the forerunner of the industrial growth, economic progress and overall development of the state. Apart from meeting industrial requirements, it also supplied electricity to domestic, municipal and township needs catering to many parts of Orissa (Murty and Mishra 1994: 132–133).

However, the study has shown that the dam remained far away to fulfil its target in terms of hydropower generation. In the year 1956, Hirakud dam started its hydropower generation with the installed capacity of 123 MW and its production significantly increased up to 1964. Additional installation capacity of 147 MW was raised in its second phase in the year 1964; however, the power production did not increase significantly. Rather, there was high fluctuation in production. Further in 1990, 37.5 MW of installation capacity was raised, but there was no increase of power production at all. The trend remained the same. Instead of increase in power production, it has been reducing from the past six years and keeps fluctuating. The highest power generation of Hirakud dam is 1,307.1623 million units in the year 1990–91 and the lowest is 507.3400 million units in the year 1960–61. It has not generated the predicted 1,524 million units of electricity till today. The annual average generation of electricity is 877.13 million units which is 57.55 per cent of the prediction (Figure 1).

Figure 1

Hydropower Generation of Hirakud Dam

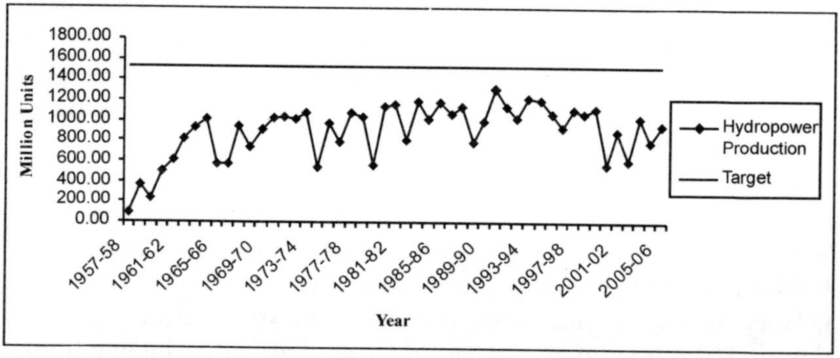

Source: Government of Orissa (2006), *Flood Report of Hirakud Reservoir for 2006*, Department of Water Resources, Superintendent Engineer Hirakud Dam Circle, Burla, p. 40.

It raises number of questions in our mind like what are the reasons for such failures and the study has shown that here are various reasons of inconsistent and under-performance of hydropower production. It was earlier mentioned that hydropower production generally depends on the water level of the reservoir. The highest water level of the Hirakud reservoir is considered as 630.00 ft on 1 November and as per the government records, it has enabled to keep its highest water level capacity (630 ft) for 99.74 per cent times in the last 49 years. But the power generation does not show any increasing trend. Rather, it has generated too less in spite of full water level in the reservoir (see Figure 2).

Figure 2

Water Level of Hirakud Reservoir (1 November)

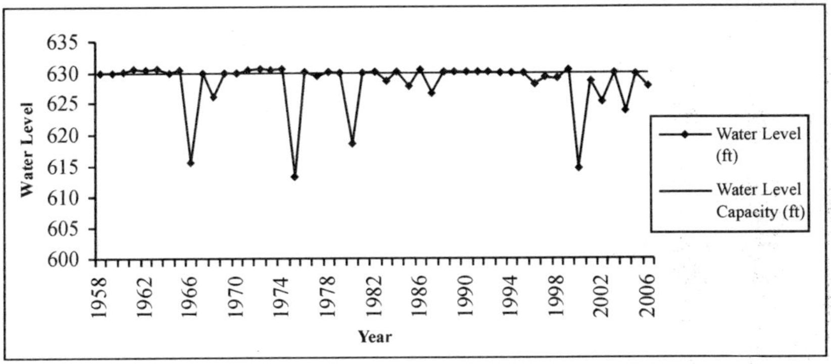

Source: Government of Orissa (2006), *Flood Report of Hirakud Reservoir for 2006*, Department of Water Resources, Superintendent Engineer Hirakud Dam Circle, Burla, p. 41.

The study has shown that such under-performance is mainly due to the breakdown of machines and turbines for renovation work from time to time. Renovation work generally gets delayed all the time due to non-availability of spare parts when required. Since the machines of the power generating units are too old and outdated, it becomes difficult for the engineers to get ready in time. Sometimes, complex administrative procedures delays the renovation and reparation work and hence, all turbines never run at a time. On the other hand, moss deposition in the mouth of power generating units too have been causing slow flow of water into the power generating units from the reservoir, which is causing another reason of low power generation, said one of the engineers at power generating units.[1]

Navigation

The proposition of navigation on the river Mahanadi was clearly stated in the feasibility report with the following objectives. It stated that, the great river Mahanadi of Orissa affords a magnificent highway for the products of Central India, but during the monsoon they became dangerous for navigation due to high floods. On the other hand, during the non-monsoon period, the volume and the current of the water get sluggish. Numerous sand banks and jutting rocks obstruct the channels in the dry season and so only small boats can be navigated. Therefore, navigation on the Mahanadi has a great future if the flow of the river can be regulated. The Mahanadi Valley and its adjoining areas are rich in mineral and forest wealth. The great water way of the Mahanadi with proper regulation arrangements would, thus, open up great possibilities of inland water transport for the distribution of industrial and agricultural products of the area in the local markets as well as for their export by sea to other parts of India and abroad.

The present port of Chandbali provides two berths for steamers about 250 ft. in length. The deep water fore-shore extends approximately 2,000 ft. upstream from the port office. The distance is sufficient for six ships of 250 ft. The capacity of the port was 800 tonnes per day and can be increased with possibility of 2,400 tonnes per day, which would increase the capacity by 40 per cent. Another site is too available, which is 16 miles downstream of Chandbali at Dhamra. The site appears to provide sufficient water to accommodate 10 ships of 300 ft. With the help of modern mechanical equipment, the capacity of the carriage can increase up to 800 to 1,000 tonnes per ship per day. The construction of other two dams; one at Tikarpara and another at Naraj would further enhance the scope of navigation of Mahanadi Valley Development Plan (Government of India 1947: 67–70).

But the plan completely failed in reality. Navigation on the river Mahanadi was neither started after the construction of Hirakud dam nor has it any hope to start in the future. As navigation is an integrated plan of Mahanadi Valley Development, it is not possible for Hirakud dam alone to fulfil this objective. The non-construction of other two dams, one at Tikarpara and another at Naraj is the ultimate factor of non-fulfillment of navigation as an objective. The full development of these facilities can be possible only after the construction of other two lower dams.

Irrigation

Although Orissa has fertile land and suitable climatic conditions to grow a rich variety of crops, it doesn't ensure successful harvest of paddy. The crop fails on average once in three to four years, primarily due to the unequal distribution of rains over the year. Rabi cultivation too gets affected due to scarcity of water and hence the rich climate and the land remained unutilized.

The Hirakud dam project was built to address this misery and it brought water through its perennial canal system to irrigate the thirsty land and promised a new life to land and people of Orissa (Agarwal 1967: 43).

The project proposed to irrigate 10,94,953 acres of land, of which 6,19,035 acres would be by flow and 4,75,918 acres by lift (Government of India 1947: 18–19) Out of this, 8,75,210 acres of land would be irrigated during *Kharif* and the rest 2,19,743 acres in *Rabi* season (ibid.: 51). In addition to 1,094,953 acres of land to be irrigated in Sambalpur and Sonepur districts, the project also proposed to irrigate in the delta area. The existed delta canals were irrigating 4,06,000 acres of cultivable area; however they were non-perennial in character, largely due to defective system of water distribution and heavy wastage of water. It too proposed 6,12,000 acres of irrigation through lift irrigation in both Cuttack and Puri districts and it ensured security of farmers income of Sambalpur, Sonepur and the delta zone (Agarwal 1967: 45–48).

Irrigation water from the Hirakud dam was first let out in 7 September 1956 and it completed 50 years in 2007. The study has shown that there is long gap between the target and actual performance of irrigation of the project. The annual average irrigation of in the command area is 6,11,580.8 acres, which fulfils only 55.85 per cent of the target. In the early years, the area of irrigation was slightly low and it is more or less constant from the year 1973. The project has irrigated maximum area of 6,87,319 acres in the year 1981–82 and the lowest 3,84,528 acres in 1966–67 (Figure 3).

Figure 3

Irrigation of Hirakud Dam (Command Area)

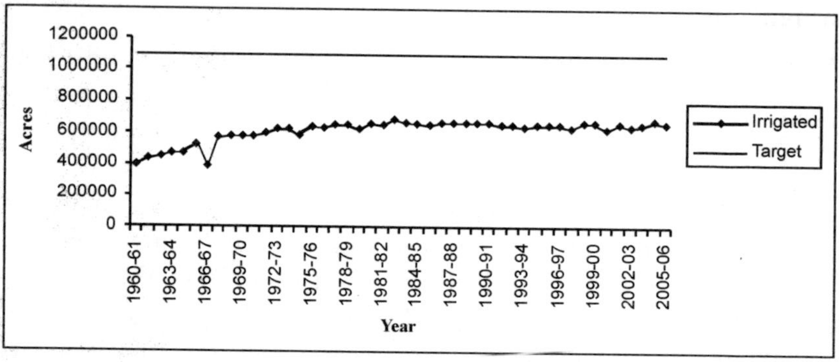

Source: Government of Orissa (2006), *Flood Report of Hirakud Reservoir for 2006*, Department of Water Resources, Superintendent Engineer Hirakud Dam Circle, Burla, p. 40.

Figure 4

Irrigation of Hirakud Dam (Kharif)

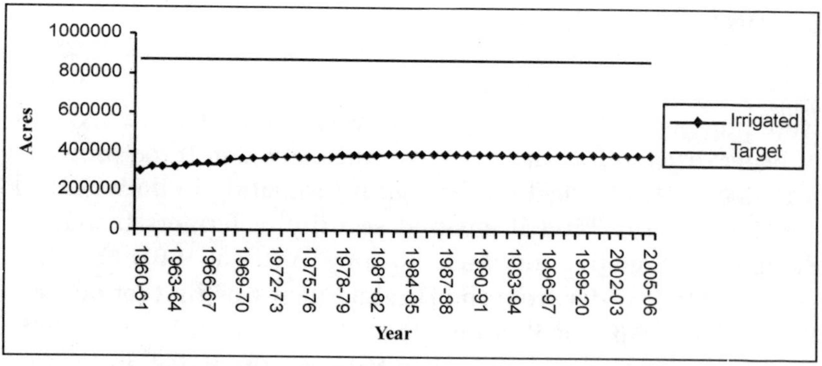

Source: Government of Orissa (2006), *Flood Report of Hirakud Reservoir for 2006*, Department of Water Resources, Superintendent Engineer Hirakud Dam Circle, Burla, p. 40.

However, the individual performance of irrigation in case of *Kharif* season and *Rabi* season is showing different picture. In case of former, the performance is too low than the target, and the latter it is adverse of the former. The annual average irrigation in the *Kharif* season is 3,76,101.5 acres, which fulfills only 42.97 per cent of its target (Figure 4). On the other hand, in case of *Rabi* season, it is exceedingly well irrigating over 2,36,566.1 acres, which is 107.15 per cent of the target (Figure 5). Although it was initially performing below its target, since 1971, it has exceeded its target. The highest performances of irrigation in case of *Kharif* (3,93,121 acres) are in the years from 1997 up till 2006. The lowest performance is 3,00,000 acres in the year of 1960–61. In case of *Rabi*, the highest performance of irrigation is 2,94,694 acres in the year 1982–83 and the lowest 44,528 acres in 1966–67.

Figure 5

Irrigation of Hirakud Dam (Rabi)

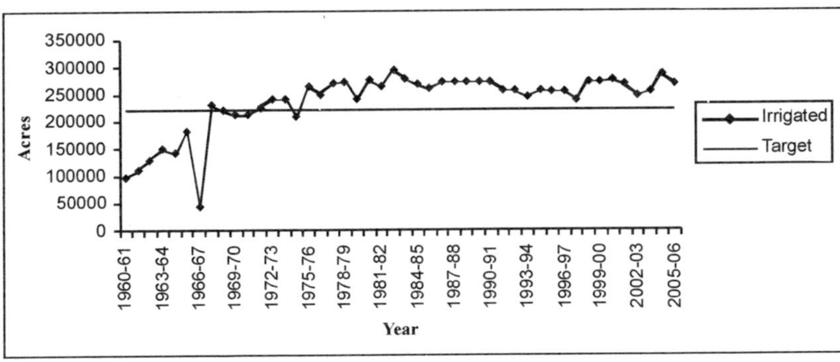

Source: Government of Orissa (2006), *Flood Report of Hirakud Reservoir for 2006*, Department of Water Resources, Superintendent Engineer Hirakud Dam Circle, Burla, p. 40.

Although the dam has been performing well as far as *Rabi* season is concerned, however the overall performance of irrigation is quite bad and, therefore, it is necessary to correlate the command area of irrigation with the water level of the reservoir. As it was seen earlier that the reservoir has maintained its full water level capacity of 99.74 per cent times in the last 49 years (Figure 5), however, the dam has not been able to irrigate the targeted acres of land during *Kharif* season in any single year. Also, the proposed 4,75,918 acres of land to be irrigated by lift was not implemented by the government. It clearly underlines that the prediction of 8,75,210 acres of land for irrigation during *Kharif* season was an overestimation by the planners.

But the argument of overestimation cannot be given as a sole factor without looking into other aspects. The governmental officials argue that the main reason of non-fulfilling of targeted area of irrigation in case of *Kharif* is changes in the distribution of crops. The assumed *Kharif* cultivation of paddy (70%), sugarcane (10%), cotton (5%) and other crops (15%) in the feasibility report has not been practiced by farmers (Table 3), who have been only cultivating paddy, which contributes 95 per cent of their total cultivation and the rest 5 per cent are other crops. Among the entire range of crops, paddy is the most water intensive cultivation and therefore the share of water released for irrigation from the reservoir is unable to irrigate the assumed area. Officials also argue that wastage of huge amount of water by farmers is also one of the major reasons for non-fulfilling of such targets. On the other hand, a Sambalpur based NGO, Manav Adhikar Seva Samiti (MASS) has found in its study that the Hirakud dam is not releasing the actual share of water proposed for irrigation. It is only releasing 8 cusecs out of shared 11 cusecs of water for irrigation (Manav Adhikar Seva Samiti 2006: 8).

Further, farmers allege that the diversion of water from the Hirakud dam to various industries located around the reservoir is one of the most important causes for decline of irrigation in the command area now a days. The government statistics, however,

Table 3

Irrigable Area of Hirakud Dam and its Crop Distributions

Kharif	*Irrigable Area (Percentage)*	*Acres*
Paddy	70%	6,12,800
Sugar cane	10%	87,470
Cotton	5%	43,735
Other Crops	15%	1,31,205
Rabi		
Wheat	25%	1,53,050
Cotton	5%	43,735
Other Rabi Crops	17.5%	22,958
Total		10,94,953

Source: Government of India (1947), *Mahanadi Valley Development: The Hirakud Dam Project,* Central Waterways Irrigation and Navigation Commission, Government of India Press, Shimla, pp. 51.

show a different picture. They show that the total area of irrigation in the command area of Hirakud dam is not declining. To be sure, there is little fluctuation, but there was no declining trend visible. Rather, in the last nine years (1998–2006), the irrigation in case of *Kharif* has been constant and highest (3,93,121 Acres). The *Rabi* is also not showing any declining trend. There is large difference between the government statistics and farmers opinion (Government of Orissa 2006: 40).

It is undeniable fact, that with the construction of Hirakud dam, it not only managed floods to certain extent in the coastal Orissa, but also brought development in the region in terms of irrigation and hydropower generation. Development of farmers of the region has accelerated to a great extent with the opening of irrigation. Industries and mining activities flourished to a large extent with the supply of hydropower. Thus literally, the hydropower system became the fore-runner of the industrial growth, economic progress and overall development of the state. Apart from meeting industrial requirements for hydropower, it also supplied electricity to domestic, municipal and township needs catering to many parts of Orissa (Murty and Mishra 1994: 132–133).

Although it brought economic progress of the state, it raises several questions. First, the failure of the Hirakud dam in fulfilling its objectives, which was envisaged in original plans (Government of India 1947: 50; Government of Orissa 2006: 132). Secondly, the large number of population displacement and their wide range of impoverishment risks with little compensations and failure to rehabilitate them by the government raises the question of equality and justice. Oustees lost their sources of income, wealth and social goods (liberty, opportunity and bases of self respect). Thirdly, it raises the issue of environmental degradation, its costs, benefits and sustainable development. Fourthly, it is a fact that Hirakud dam was made for the purpose of development and national prosperity and in this process, many people sacrificed for the sake of national interest, however, observing closely, the benefits made by the dam is leading to uneven development and strengthening capitalism, especially of large farmers and industrialists.

Extent of Displacement

Applying the LAA of 1894 for greater good of the greater number, Hirakud dam submerged vast areas of land and displaced number of people. The empirical record shows that it has submerged more lands and displaced more people than estimated in the feasibility report (Table 4). It was estimated in the feasibility report that about 168 villages would be submerged covering 1,35,000 acres of land under the Hirakud dam reservoir. Out of which 70,000 acres will be cultivated land (Government of India 1947: 315–317). But after the construction, it submerged 325 villages covering 1,83,000 acres of land. The submerged cultivated land was estimated at 1,23,000 acres of land. It submerged 291 villages in Orissa and 34 villages in Madhya Pradesh and displaced about 26,501 families (approximately 1,00,000 people) (Government of Orissa 2007: 1).

Table 4

Proposed Estimation in 1947 and Actual in 1957 after the Completion of Hirakud Dam

	Proposed (1947)	*Actual (1957)*
Submergence of Total Land	1,35,000 Acres	1,83,000 Acres
Submergence of Cultivable Land	70,000 Acres	1,23,000 Acres
Number of Families Displaced	–	26,501
Number of Villages submerged	168	325
Number of Police Stations Submerged	08	09

Source: Government of India (1947), *Mahanadi Valley Development: The Hirakud Dam Project*, Central Waterways Irrigation and Navigation Commission, Government of India Press, Simla, pp. 315–317 and Government of Orissa (2007), *Hirakud Cell Status Report* (HLO), Collectorate Office, April, Sambalpur, p.1.

Villages acquired for submergence of the Hirakud dam can be classified into three categories (completely submerged villages, partially affected villages and villages acquired beyond 632 R.L.). It completely submerged 116 villages, partially affected 133 villages and 42 villages acquired beyond 632 R.L. (Reservoir Level).[2] Among those 133 partially affected villages, 48 villages where *basti* sites or the houses submerged, but the cultivable land and some houses remained outside. Rest 85 villages where land have been submerged and Basti-site or the houses remained outside. The pattern of submergence of 34 villages in Madhya Pradesh is not mentioned here due to lack of information, unavailable at concern authority of Orissa (Table 5) (Government of Orissa 2007).

Table 5

List of Villages Acquired for the Submergence in the Project Hirakud Dam in Orissa

Name of the Police Station	*No. of Completely Acquired Villages*	*No. of Villages where basti sites submerged but cultivable land and some houses remained outside*	*No. of Villages where only land have been submerged*	*No. of Villages where land acquired beyond 632 R.L.*	*Total*
Rampella	23	07	08	15	53
Mura	27	11	07	07	52
Mahadevpali	19	04	23	04	50
Attabira	12	0	0	01	13
Ambhabhona	12	19	19	07	42
Jharsuguda	02	17	17	03	30
Katarbaga	05	10	10	05	29
Sason	01	01	01	0	04
Sadar	15	0	0	0	18
	116	48	85	42	291

Source: Government of Orissa (2007), *Hirakud Cell Status Report* (HLO), Collectoriate Office, April, Sambalpur.

Resettlement and Rehabilitation

A new department known as 'Hirakud Land Organization' (HLO) was set up by the state government of Orissa for orderly evacuation, resettlement and rehabilitation of the displaced people. There were three officers working under the HLO. Such are (i) the Land Acquisition Officer (ii) the Land Reclamation Officer and (iii) and the Settlement Officer (Agarwal 1967: 71–72).

Land Based Strategy: The displaced people were rehabilitated in 18 different camps (Table 6). It resettled only 2,243

Table 6

Details about Rehabilitation Camps for the Displaced by the Hirakud Dam

Name of the Block of the Resettlement Camp	Name of the Resettlement Camp	Name of the Connected Revenue Villages of Resettled Camp	No. of Families Settled in the Camp	No. of Families Living	No. of Outsiders
Attabira	Attabira Colony	Nua Batimunda	82	82	0
Jujumura	Basipada 'A'	Birsinghgarh	49	49	61
Jujumura	Basipada 'B'	Nua-Mahulpali	61	61	0
Jujumura	Badasahir	Nua-Barangmal	386	386	0
Jujumura	Chhamunda	Nua-Adapada	201	277	0
Maneswar	Hotapali	Larasara, Deogaon	0	0	92
Dhankauda	Nua-Hirakud	Nua-Hirakud	39	38	0
Dhankuda	Lamdungri	Lamdungri	83	83	0
Dhankuda	Sangarmal 'A'	Nua-Mura	309	309	14
Dhankauda	Sangarmal 'B'	Tihura	265	265	14
Rengali	Jharghati	Nua-Rampella	310	298	0
Rengali	Jhranpali	Nua-Tiligi	25	25	0

Cont'd ...

... Cont'd

Rengali	Sardapali	Lpanga	8	8	2
Rengali	Laumal	Nua-Khurigaon	182	182	0
Jamankira	Goudpali 'A'	Kadalipal	54	12	13
Jamankira	Goudpali 'B'	Nua-Bhoghara	36	27	9
Jmankira	San-Kundesara	Nua-Burda	153	140	0
Kuchinda	Udsingh	0	0	0	0
Total			2,243	2,242	205

Source: Government of Orissa (2007), *Hirakud Cell Status Report* (HLO), Collectoriate Office, April, Sambalpur, pp. 13–15.

displaced families, which merely constitutes 8.46 per cent of the total displaced. Others were resettled themselves by their own effort (Government of Orissa 2007: 1). The reason for this are varied. Some felt that these camps were too far away from their original villages, located in the densely forested area, and inadequate in terms of infrastructural development. Various people found difficulties in shifting their luggage (tile, bricks, wood and cattle etc) to those rehabilitation camps. Others had no faith on the government scheme and found the uncertainty of the new locations. Thus, many people took advantage of the hospitality of neighbouring villages and opted to settle there (Viegas 1992: 49). Also, many of them preferred to stay in the periphery of Hirakud dam reservoir, now practicing cultivation on the excess acquired land on temporary lease basis. It was proposed in the feasibility report that 50 per cent of the displaced would be rehabilitated, but such faulty resettlement planning detracted them to resettle in camps (Government of Orissa 1989: 4–5). It is clearly visible that even now two of the rehabilitation camps Udsing and Hotapali are still vacant. Among those, the latter is acquired by outsiders. Besides Hotapali camp, outsiders are also staying in other six camps (Basipada 'A', Sangarmal 'A', Sangarmal 'B', Sardhapali, Goudpali 'A', Goudpali 'B') where lands were laid vacant. In total, there are 205 outsider families have been staying in seven resettlement camps, including Hotapali (Government of Orissa 2007: 13–15).

The oustees, who really wanted to be rehabilitated by the government, were just dumped in those identified camps (full of jungles and wild animals), which were devoid of basic amenities of life and without basic sources of economy, roads, schools, hospitals, clubs, markets, etc. It has been shown in my study (Kadalipal) that the government did not provide any land for cultivation to the oustees. They were only given meagre compensations for their submerged lands.

The process of compensation started on 14 April 1949 by the HLO working under the supervision of Revenue Divisional Commissioner (RDC). However, the programme was discontinued in the year 1978 (Government of Orissa 2007: 3). The rates of compensations for likely submerged 'lands' proposed in the feasibility report ranged from ₹ 50 to ₹ 1,000 per acre according to their classification in terms of productivity.[3] But in actual, they were paid more or less in a uniform rate ranging from ₹ 200 to ₹ 600 per acre, which was much less than the market value (Government of India 1947: 315–317). Similarly, 'houses' were also classified into seven categories[4] and the rate of compensations ranged from ₹ 250 to ₹ 3,000. Even, the oustees got the exact amount as per the provision, but it was too less to construct a new home elsewhere (ibid.: 318–319). For 'temples', the rate was estimated at ₹ 3,000 and ₹ 1,000 for *pucca* (concrete) and *Kachcha* (made of mud) temples respectively, but no compensations were made in any of the categories, largely due to the fragmented and distorted resettlement pattern of displaced people. It became problematic for the government to whom it would offer the amount of compensations to construct the same (ibid.: 319). The rate of compensations for 'schools' and 'tanks' was estimated at ₹ 1,000 for each of those, and in such cases, similar problems also occurred (besides privately owned tanks and wells) like temples. Even if, compensations were made for privately owned tanks, it was not given as per the amount mentioned in the feasibility report; rather those were simply considered as lands and compensated accordingly. In case of 'wells', people got compensations as per the provision (₹ 250 and ₹ 60 for 'masonry' and 'Kachcha' wells respectively), but it was too less (ibid.: 319–320). Value of 'trees' was estimated, varying from ₹ 10 to ₹ 30 for fruit bearing trees and ₹ 10 for timber producing trees, but it has not been followed. In many cases, conflicts arose regarding the ownership of trees and at last the third party (mostly influential people of villages, largely village head) got the benefit of compensations. The *Gountias* (village heads) had considerable influence in the society as well as over government officials during that period and they managed to grab compensations as they could by *making fool of common people*[5] (ibid.: 320).

The absence of land for land as compensation was a great blow to the oustees and it was very difficult to buy a piece of cultivable land elsewhere with the meagre compensations they received. Therefore, the forests reclamation activities were made by the oustees themselves to convert the forest land into cultivated land. The government even did not provide the expenditures of land reclamation activities and, therefore, it was a huge burden on the resettled, who were already in a situation of severe hardships. It is too ironical that many of them were not given the legal ownership of those lands by the government. Now, the government is claiming that these lands are forest land, coming under the public purpose of roads, drainage and water tanks, etc. Thus, a sense of insecurity still persists among them to secure their livelihoods. Besides it, a number of fishing communities, earlier dependent on the river Mahanadi for fishing to maintain their livelihood, were forced to leave the camp, since they did not find the place suitable for their source of livelihoods.

The camp too was devoid of basic facilities of life like school, hospital market, club etc and the government did not provide any such facilities later. However, the resettled people opened up a school by their own initiatives; but it is ironical that the school was not given recognition by the government, since the rehabilitation camp was not recognized as revenue village for longer period of time. The school got recognized by the government in the year 2004. The absence of the title of revenue village further victimized them on other regards as well. They were deprived of all the benefits of several welfare programmes undertaken by the government from time to time and, thus, remained marginalized. Such situations prompted many families to leave rehabilitation camps to settle elsewhere and the camp has recorded the highest number of families leaving the camp. Taking the opportunity of emptiness of camps, many outsiders now have been staying in the rehabilitation camp, which has further caused the feeling of insecurity in the minds of the families staying in the camp. Interestingly, most of the outsiders are from the state of Bihar. The questions regarding the probability of land grabbing by outsiders and resource constraint and that it might cause conflicts in the future are a source of worry for the rehabilitees.

On the other hand, my survey in the resettled village (Jammal) has shown that most of the oustees resettled voluntarily in that village are belong to Scheduled Tribes (STs). Since they had no legal title over their lands, where they used to cultivate and maintain their livelihoods, these lands are declared as public lands by the government and, therefore, they were deprived of any compensations of their submerged lands. In case of non-tribal oustees, however, they were given little compensations for their lost assets and properties, which was very much similar to the experiences made in resettlement camp of Kadalipal.

It was found from the study that especially the displaced tribals, resettled in Jammal village, were forced to work as wage labourers. Earlier, they were considered as masters of their own. Deprived of compensations and any kind of help from the government, they were reeling under severe socio-economic hardships. With the shift of accustomed environment and ecology, the displaced people faced severe 'health crisis'. Among all diseases, Malaria was very much rampant among them; primarily the region was densely covered with forests. Many people died of Malaria owing to the lack of medical facilities. It is true that the non-displaced people of the region were also suffering from various diseases and dying of Malaria, but the intensity of the health problems was more in case of the displaced people, since they were newly resettled in the new region. Also, many people died out of their unbearable psychological trauma. The severity of socio-economic impacts caused by displacement was even more on 'women' than men. Women in general in this region were engaged in their household activities; however, in their newly resettled village, they were forced to work outside to eke out their livelihood. The children are the worst victims, who were forced to discontinue their studies and remain deprived of education and to fetch a job in the future to secure their livelihoods. The displaced too were deprived of several benefits of governmental welfare schemes for several years, since they were not the original resident of that village. The government took long time to recognize them as resident of that village.

Besides tribals, there were a number of non-farming communities (barbers, car painters and blacksmiths), village priests and village guards who were also resettled in that village, who were earlier maintaining their livelihood by practicing their traditional occupation from generations. However, with their displacement, they lost their clients as well as their occupation and livelihoods. The same problems were faced by the petty businessmen as well. But the problems of village priests and village guards are little bit different from the non-farming communities. They had quite sufficient amount of lands, but did not get any compensation from the government. Because the lands belonging to them were gifted lands by the villagers for their services. The absence of legal ownership of such gifted lands deprived them of compensations and therefore livelihood crisis persisted among them[6].

Non-land Based Strategy: The government had the plan of resettlement and rehabilitation programme for the oustees based on land-based strategy, mentioned in the feasibility report. However, neither there was provision with regard to non-land based strategy nor the government has initiated such programmes later to raise the sources of livelihoods for the displaced.

Impoverishment Risk

The study has shown that the oustees have reeled under severe 'impoverishment risk'. Evidences obtained from the oustees show that the submergence of their land affected severely in terms of their loss of livelihoods and income sources. They also lost their arable land, common property resources (forests, grazing land, ground and surface water, fisheries and so on), which changed the access and control of productive resources. The loss of economic power with the breakdown of complex livelihood systems led to decline of their living standards and marginalized them in the society. On the other hand, the loss of livelihood and disruption of agricultural activity of the oustees adversely affected their household food security. The displaced also faced higher incidence of diseases associated with new environment and ecology and many died out of mental trauma, which increased morbidity and mortality. The involuntary and forced displacement also fragmented their existed social fabric, which led to socio-cultural disarticulation.

Question of Justice

No doubt Hirakud dam has brought massive economic progress of the state and has become the fore-runner of industrial growth. It contributed a lot in managing floods in coastal Orissa and provided electricity throughout the state to both industrial and domestic sectors. In addition, it brought irrigation to quench the thirsty land of farmers of western Orissa, and it raised their incomes. However, it is ironical that none of the oustees got any of the benefits of the project. Rather, they were deprived in the name of 'greater good of the greater numbers'. The oustees were not given even a piece of land in the irrigated area as compensations to maintain their livelihood. They were also not given electricity produced by the dam and most of it is supplied to industrial, mining and urban sectors. On the other hand, the unsuccessful resettlement and rehabilitation of the oustees created a new inequality by depriving them not only of 'income and wealth' but also of social goods in two other ways: 'liberty and opportunity' and 'social bases of respect'. The liberty of the oustees was violated, since their displacement was involuntary and forced by the government of Orissa without any consultations to the victims. The social disruptions caused by the dam deprived oustees from their self-respects and put them in the new scheme of compensations, resettlement and rehabilitation. According to Rawl, the scheme or system, which creates advantage for some parts of community, is unjust, unless it creates benefits for all. Hence, the project Hirakud which is creating advantage to some and depriving others is unjust.

Ecological Changes and Environmental Impacts

Over half a century, after the construction of the dam, its catchment, reservoir and command have undergone considerable economic, environmental and ecological changes. These changes have significantly affected water use and availability, both in terms of quality and quantity, as well as inflows into and outflows from the dam. The rainfall pattern in the region too has changed considerably and, therefore, impacted a lot to farmers. Looking at the above scenario, management have been changing its priorities of natural resources in these different spaces and, therefore, it has been putting strong impacts on the evolution and manifestation of the water conflicts around Hirakud (Choudhury, Sandbhor and Satapathy 2012: 29).

As far as ecological changes are concerned, the Hirakud dam submerged 1,600 hectare of reserve forests and 20,400 hectare of village forests. The loss of forests or deforestation activities took place further by industrial houses in the Mahanadi catchments, soon after the developmental plan periods came into force. Study made by A. Mishra and M.C. Dash (1984) says that deforestation is found to be an accelerating process where the rate of loss is estimated to be more during the period 1970–1975 and 1960–1970 in comparison to 1950–1960. It is largely due to various developmental plans that came into force after 1960s. They argue that the vast submergence of forests land has brought massive ecological changes, heavily impacted on the pattern of rainfall in the locality. It has been testified that the general evaporation rate has increased to a great extend, causing desertification in the western Orissa. After the construction of the dam, in 1965, Sambalpur tasted the bitterness of famine. It has made the Bargarh a rain shadow area, where there are no Mangroves or forests. The nature of climate change has put its adverse impacts on the rainfall pattern in the region. The folklore (*Varsa Chaturmasya*) of that locality says that there were about eight days of heavy rain, 16 days of moderate rain, 32 days of slow rain and 64 days of drizzle for healthy crops. But now this has come down to 50 days. It has been recorded in the original project report of the Hirakud Reservoir (1947) that the average annual rainfall of Sambalpur was 1381mm and it has reduced to 1369mm in the year 1953, as mentioned in its second revised project report. If we see the average monsoon rainfall in the post Hirakud period from 1959 to 2006, it has come down to 1142mm, and have been affecting severely the local cultivation. Besides, the reducing rainfall and its irregularities, people too do not feel the earlier cold waves of winter in the local area, since they do not find dew drops on account of massive ecological changes (Mishra and Dash 1984: 51–58; Pati and Biswal 2009: 9–10).

According to the preliminary consolidation report (2008) on climate change by the Central Water Commission (CWC), the Mahanadi river basin is experiencing an increase in mean annual temperature of 0.40 to 0.60°C every 100 years. Seasonal analysis shows that there is an increase in rainfall in the post-monsoon period and experiencing the maximum decrease in the number of rainy days in India. In the last 100 years, the decrease of annual rainy days in basin is 4.70 per cent and there is 10.53 per cent increase in post-monsoon rainy days. On the other hand, there is an increase of 16.60 mm in heaviest rainfall spell of the year in the last 100 years. The report of Intergovernmental Panel on Climate Change (2007) says that the Mahanadi river system falls in a tropical region which is going to see more floods, a decrease in the average runoff, and frequent droughts in the future on account of climate change. The water availability in the region will be under stress (cited in Choudhury Sandbhor and Satapathy 2012: 33–34).

The developmental activities in the catchment area too have affected significantly the water availability, its use, both in terms of quality and quantity, as well as inflows into and outflows from the dam. The catchment of the Hirakud dam extends over an area of 83,400 sq km, which is 58.9 per cent of the total catchment area of the Mahanadi basin, lies in the states of Chhattisgarh (74,970 sq km – 89.9%) and Odisha (7,400 sq km – 8.8%) with small portions lying in Madhya Pradesh (130 sq km), Maharashtra (250 sq km) and Jharkhand (650 sq km). The inflows into the Hirakud reservoir are mainly contributed by six upstream sub-basins: Upper Mahanadi, Seonath, Hasdeo, Jonk, Mand and Ib. After 1957, the Government of Madhya Pradesh constructed a number of dams in the Mahanadi basin for the purpose of drinking water, irrigation and industrial water supply, etc. without inter-state agreement with the Government of Orissa. In a meeting held at Pachmarhi on 15 June 1973, the Government of Madhya Pradesh argued that it had full rights to irrigate at least 50 per cent of cultivable area of the Mahanadi basin in the Chhattisgarh plains and accordingly, a number of major, medium and minor water resource projects have come up in the upper catchment, leading to a changed inflow pattern at the dam site. Apart from five major and medium projects, about 150 small dams have been constructed on the Mahanadi and its tributaries upstream of the Hirakud reservoir in the past 50 years. Twenty eight tanks have been constructed in four sub-basins in Chhattisgarh alone, as of the year 2009, and are intercepting flows from a 16,844.94 sq km catchment area, out of a total 83,400.00 sq km catchment area of the Hirakud reservoir. A number of water guzzling industries too have now signed Memorandum of Understanding (MOUs) with the Government of Chhattisgarh; have been allocated more than 1,035 MCM of water (ibid.: 29–33).

Further, the increase in area under watershed treatments and water harvesting structures, especially in Chhattisgarh, and the reduction in forest area in the catchment over the last 50 years have too changed the catchment characteristics considerably, which has significantly affected inflow quantity, seasonality, as well as silt yield in the reservoir. The Government of Chhattisgarh has introduced a number of watershed development programmes to accelerate intensive cultivation in the state especially after the 1990. About, 8,49,154.89 hectares of land covering 3,771 villages is treated under the Drought Prone Areas Programme (DPAP), Employment Assurance Scheme (EAS) and watershed development programmes. On the other hand, another 175 watersheds covering an area of 18.5 million hectares area is treated under the National Watershed Development Programme for Rainfed Areas (NWDPRA) (ibid.: 33).

The increasing entry of industries and several developmental activities in the catchments has not only affected the quantity of flow, but also the quality of water. The total industrial effluent released from different industries into the Mahanadi at Sambalpur is 736 KLD (Kilo Litres per Day) and discharge 1,00,000 cubic metre of waste water every day, and has been causing harmful effects on the people of these areas, their livestock, and cultivated crops. It has totally ravaged the Sambalpur-Jharsuguda region rapidly into the largest conglomerate of smelter and power industrial units in the world. The increasing polluted water is no longer supporting the traditional occupation of agriculture and therefore forcing people of another means of livelihood (ibid.: 27).

A variety of industries have been established in the region owing to the vast mineral and power generation infrastructure. The important industrial regions in the basin in Chhattisgarh are Bhilai, Durg, Raipur, Rajnandgaon, Korba, Urla and Bilaspur.

The iron and steel industry at Bhilai, cement industries at Durg and Raipur, textile industry of Rajnandgaon, and aluminium and thermal power plants at Korba are the major polluting industries in Chhattisgarh, which drain their effluents in the river basin. In Orissa, mining activities by Mahanadi Coalfield Limited (MCL) in the Ib valley and Lakhanpur block of Jharsuguda, open cast mines at Brjarajnagar and Belpahar, which directly release effluents into the river stream. According to the Comprehensive Environment Pollution Index, out of 88 polluted industrial areas, the Ib Valley and Jharsuguda are among the critically polluted area. Study made by the National Environmental Engineering Research Institute (NEERI) says that the most toxic effluents generating industries are in the Rengali and Jharsuguda areas. More than 3000 MTP of fly ash from the industries directly flows into the water sources, which ultimately join the Hirakud reservoir and among them, Vedanta project bears the maximum level of fluoride, i.e. 6.5 mg per litre. According to the Odisha State Pollution Control Board, the present rate of solid waste generation due to industrial activities in Sambalpur and Jharsuguda region is 4.58 Million Tons Per Annum (MTPA), and is expected to increase almost 13 times to 58 MTPA by 2031 which would require 13,000 ha of land in the peripheral region of the reservoir for its disposal. It has further stated that the ongoing process of industrialization in the region and its future plan would annually consume about 80,000 tonnes of fluoride-bearing materials and considering it, about 800 tonnes of these materials may be washed into the reservoir. It has raised an alarm that if the Hirakud reservoir is contaminated with fluoride, there might be an epidemic of fluorosis in the region. A study on water quality by the Central Pollution Control Board (CPCB) says that the Mahanadi is less polluted compared to other important rivers in the country. However, certain stretches have a comparatively higher degree of pollution. Korba has been identified as a critically polluted area in the basin. The industrial as well as domestic wastewater is being discharged directly into the Hasdeo and Ahiran rivers, as well as the Dengur nala. The major source of pollution in the basin is thermal power plants, the Bharat Aluminium Company, the captive power plant of

Bharat Aluminium Co. Ltd., and coal mining, most of which are located on the banks of the Mahanadi River or its tributaries and distributaries (ibid.: 34–35).

Industrial effluents coming out from industries are not only detrimental to human beings, but also it has been severely affecting standing crops. In the year 2011, the crops of about 35 villages in Sambalpur, Jharsuguda, Bargarh and Sonepur districts were affected on account of fluoride contamination of the flood water coming from the Hirakud reservoir. The crops were destroyed and burdened them of heavy economic loss. The study carried out by the State Pollution Control Board has indicated that with the ongoing industrialization in the Jharsuguda-Sambalpur region at the periphery of Hirakud would generate about 1,26,000 tonnes of fluoride bearing hazardous liquid waste annually, which puts the reservoir under a significant risk of fluoride contamination. This perspective of pollution due to floods, in terms of the additional damage in non-traditional areas near the reservoir, is a new dimension which links the reservoir route flood with the industry-agriculture water conflict (ibid.: 17).

Uneven Development

Although, there has been a long gap between the target and actual performance in terms of irrigation made by Hirakud dam, only fulfilling 55.85 per cent of the target, it has brought massive changes in agricultural sector of the state by raising the economy and livelihood of farmers, especially of that region. Now, the command area, especially the Bargarh is known as the rice bowl of Orissa. In spite of its huge achievement of Hirakud dam, questions have been raised for its uneven development in the command area. Study made by Ahmedabad based 'Development Support System' says that there is uneven distribution of irrigation water leading to uneven development in the region. It states that all lands are irrigated in the Head Reach Area (HRA), and on the other hand, only 35 per cent of the lands are irrigated in the Middle Reach Area (MRA) and only 18 per cent in the Tail End Area (TEA) (Pati and Biswal 2009: 8).

Further, a study made by Meher (2011) states that, although in the initial years the dam was irrigating the whole command area, however since 1990s, the area of irrigation gradually started to decrease and have been affecting the livelihood of the marginalized populations. The small and marginal farmers comprising mostly of the tribals and other marginalized people living at the tail end of the command area are now deprived of irrigation water. They are not in a position to grow any second crop during *Rabi* season (winter and summer crop) due to acute water scarcity. Also, during drought years they do not get adequate water to protect their standing paddy crop in the *Kharif* (rainy and autumn crop) season, which is a timing of pressing need. He claims that although the command area of the Hirakud dam became agriculturally more prosperous, especially in its post-irrigation era, however, it again started its rapid deterioration after the 1970s. Now a days, many a farmers in the Hirakud command area are deprived of irrigation water to grow a second paddy crop during the *Rabi* season. This situation is conspicuously visible in the tail end villages of the different distributaries, minors and sub-minors including water courses at different points of the main Bargarh Canal. This has further taken on an acute form in the tail end canal of the main Bargarh Canal, which is called Bhimtikra Main Canal in the Binka block of Sonepur district (Meher 2011: 1–17).

However, the study shows that, there is a wide variation of deprivation of irrigation water in the different reaches of the command area. It is found that, in case of Bhimtikra Canal, head reach, middle reach and the tail end are equally deprived of irrigation water during both the *Kharif* and *Rabi* seasons. On the other side, in case of Bargarh Canal, the deprivation of irrigation water in the far tail end area is 100 per cent for the second crop 64.44 per cent in the middle reach area and only 10.97 per cent in the head reach area. However, for the entire tail end area of the Bargarh Canal, the degree of deprivation during the period of second crop is as high as 68.05 per cent of the stipulated *Rabi* crop area and for the *Kharif* it is only 9.93 per cent. According to the one estimate, about 50,000 acres (20,000 ha) of land in the command area of the Hirakud dam at present do not receive water to grow a second paddy crop. Hence, conflicts have been arising among the farmers of head reach, middle reach and tail end area and many a times it is leading to family, caste and village level conflicts, resulting in court cases and social and political polarizations (ibid.).

In such scenario, neither there is attempt to enhance the potential of water availability for irrigation from the government side nor the practice of crop rotation or crop diversification from farmer's side. Further, the cultivation of high yielding crops (water intensive) in the tail end area has been increasingly depriving irrigation water over the years. On the other hand, the cost of labour, fertilizers, pesticides and water taxes have been rising every year and for the last six to seven years, farmers in the Hirakud Command area have found it very difficult to sell paddy at a remunerative price. The increased cost of living and declining farming income have put the farmer in a very disadvantageous position. There are now many cases of suicides and self-immolation and a sense of helplessness is prevailing among them. Many small and marginal farmers, including tenant farmers in the Hirakud command area, borrow heavily from commercial banks and cooperatives and from private money lenders/traders at a higher rate of interests in order to meet the increasing costs of inputs such as seeds, fertilizers, hired labour and so on. In such situation, if crops fail either due to lack of irrigation water or natural calamities like hailstorms, unseasonal rains, etc. it ravages the economic foundation of farmers. They do not see any hope of coping with the problem of poverty and an increasing debt burden and it has accelerated to a great extent in the post-reform years of economic liberalization and globalization. The terms of trade between agriculture and industry have worsened and there has been a massive cut in agricultural subsidies for inputs like fertilizers, pesticides, water and power supply. Further, the rise of labour-hiring costs again has been causing the decline of agricultural output (ibid.).

It has also been found that water availability for agriculture in the future will be reduced, as the area around the Ib valley surrounding the Hirakud reservoir is now witnessing a rapid industrial growth. It has been seen that the government of Orissa has permitted several industries to take water from the Hirakud reservoir for their industrial purpose, which is 27 times more than what it was before 1997. In such a situation, the farmers at the tail end of the Hirakud command area are likely to suffer more and more (ibid.).

Agriculture Versus Industries

In Chapter 4, various types of conflicts and controversies relating to water have been discussed in detail. Most conflicts are inter-state in nature, dealing with the issue of submergence of land, environmental degradations, large scale displacements, allocation of resources or sharing of costs and benefits. However, water conflicts of Hirakud dam in the recent years are inter-sectoral in nature. Conflicts between farmers (agricultural sector) and industries (industrial sector) emerged, while the Government of Orissa gave permission to number of industries to draw huge amount of water from the Hirakud reservoir, and it decreased the area of irrigation in the tail end area, affecting the livelihood of farmer community at large.

With the construction of Hirakud dam in 1957, various industries were established in the adjacent area of dam in order to exploit the huge deposited mineral resources of the region. This was primarily due to availability of cheap electricity and easy access to water from the dam to run these industries. A number of industries have been drawing water from the Hirakud reservoir in the first phase since 1957 (Table 7). Even if, the dam had no stated objective of giving water to industries, the government of Orissa had granted permission to six industries (IB Thermal Power Station (M/s OPGC Ltd) at Bandharpali (Jharsuguda) for production of electricity, B.I.T. M.C.L at Bandabahal (Jharsuguda) for extraction of coal, Orient Paper Mill at Brajarajnagar for run its mill, S.R.L.Y. Jharsuguda for water requirement of railways and M/S INDAL at Hirakud for production of Aluminum) to draw water directly from the Hirakud reservoir for various industrial purposes. They have been collectively drawing 30.38 cusecs or 0.022 Million Acre cubic feet (MAcft) of water from the reservoir. Further, the process of liberalization in 1990s brought large scale industries and mining projects in the adjacent areas of Hirakud dam, and again, the Government of Orissa granted permission to draw 330.48 cusecs or 0.239 MAcft of water to these industries (IB Thermal Power Station (M/S OPGC LTD) at Bandharpali (Jharsuguda) for production of electricity, M/S Bhusan Ltd at Thelkoli, Lapanga and Jharsuguda for production of steel, M/S S.M.C. Power Generation Corporation at Himra for generation of electricity, M/S Sterlite Opticals Technology Ltd, at Jharsuguda for Generation of Electricity/Aluminum, M/S Aditya Aluminum at Lapanga and Jharsuguda for production of Aluminum, and Shyam D.R.I. Ltd at Rengali for production of Sponge Iron). One cusecs of water could irrigate 100 acres of land, and by that calculation farmers argue that withdrawal of 30.38 cusecs of water by industries in the first phase already deprived 3,038 acres of land of irrigation. In the second phase, 330.48 cusecs or 0.239 MAcft of water withdrawal could further would affect 33,048 acres of land (Table 8) (Government of Orissa 2007).

Table 7

Drawal of Water from the Hirakud Reservoir in the First Phase (1957)

Name of Industries	*Purpose*	*Location*	*Average Drawl of Water in Cusecs*	*Average Drawl of Water in MAcFt*
IB Thermal Power Station (M/S OPGC LTD)	Production of Electricity	Bandharpali (Jharsuguda)	12.36	0.009
B.I.T. M.C.L Bandabahal	For Extraction of Coal	Bandharpali (Jharsuguda)	3.71	0.003
Orient Paper Mill	Factory Locked, Drinking Purpose	Brajarajnagar	1.84	0.001
S.R.L.Y. Jharsuguda	Water Requirement of Railways	Jharsuguda	2.47	0.002
M/S INDAL Hirakud	Production of Aluminium	Hirakud	10.00	0.007
Total			30.38	0.022

Source: Government of Orissa (2007), Office of the Executive Engineer, Main Dam Division Burla.

Table 8

Industrial Units Granted Permission to Draw Water from the Hirakud Reservoir since 1990s

Name of Industries	*Purpose*	*Location*	*Average Drawl of Water in Cusecs*	*Average Drawl of Water in MAcFt*
IB Thermal Power Station (M/S OPGC LTD)	Production of Electricity	Bandharpali (Jharsuguda)	53.00	0.038
M/S Bhusan Ltd	Production of Steel	Thelkoli, Lapanga, Jharsuguda	100.00	0.072
M/S S.M.C. Power Generation Corporation	Generation of Electricity	Hirma	2.45	0.002
M/S Sterlite Opticals Technology Ltd	Generation of Electricity/ Aluminum	Jharsuguda	111.30	0.081
M/S Aditya Aluminum	Production of Aluminum	Lapanga, Jharsuguda	52.73	0.038
Shyam D.R.I. Ltd	Production of Sponge Iron	Rengali	11.00	0.008
Total			330.48	0.239

N.B: One cusecs of water irrigate 100 acres of land.

Source: Government of Orissa (2007), Office of the Executive Engineer, Main Dam Division Burla.

Water is a basic human need and a precious national asset. Therefore, planning, development and management of water resources is very much essential from a national perspective. The National Water Policy was first formulated in 1987 and on its basis, the Orissa State Water Policy was formulated in 1994. The National Water Policy–1987 has been reviewed, updated and a new policy titled National Water Policy–2002 has been adopted by the Government of India to address new issues and challenges that have emerged in the field of development and management of water resources. It was also, therefore, felt necessary for the

Government of Orissa to review the State Water Policy–1994. After due consideration, a new policy was prepared called 'Orissa State Water Policy–2007', keeping with the National Water Policy – 2002. It aims at laying down the principles of equitable and judicious use of water for survival of life, welfare of human beings and sustained as well as balanced growth of the state. The state of Orissa in its policy has adopted the following order of priority in water allocation in tune with the National Policy: drinking water, ecology, irrigation, hydropower, industries and navigation (Government of Orissa 2007a).

However, in practice, the Government of Orissa has not been following the principle of water allocation as per the Orissa State Water Policy–2007. In spite of giving first priority to irrigation and agriculture, the Government of Orissa has been giving utmost importance to industries. It is being argued that the economic value of water in terms of industry is much higher in comparison to agriculture. This calculation, the rationale goes, has prompted the Government of Orissa to make agreements with industries to draw water from Hirakud reservoir. This policy of laying more emphasis on industries and underestimation to irrigation and agriculture has caused deep anguish among farmers, who are the main economic stay of the people of Orissa. To protest against these policies of the state government, on 19 June 2006 the Hirakhand Nagarik Parishad[7] (Sambalpur) requested to the President of India to appoint a team of experts to study the loss of irrigation and power generation by providing water for industrial use from the Hirakud reservoir (Hirakhand Nagarik Parishad 2007: 72). The President considered the problem and directed the Department of Water Resource, Government of Orissa, to send a technical expert team to investigate the matter. On the advice of the President, the Department sent a team of technical experts headed by its Engineer-in-Chief. The team surveyed the Hirakud dam, held discussions with members of Hirakhand Nagarik Parishad – namely, Er. K.K Supakar, Bhagbat Prasad Nanda, Dilip Kumar Padhi and Er. R.C. Pujari – and then submitted the report. To the disappointment of the farmers, the report concluded that the irrigation potential of Hirakud reservoir would not be hampered, if 0.239 MAcft of water were be provided to industries. It further stated that the reduction of irrigation potential can be avoided, if proper management of irrigation water is done by reducing different losses, by lining of main canals and other improvement works (Government of Orissa 2006a).

Discontent Among Farmers

Irrigation through dam water provides for the livelihood of thousands of farmers in Orissa. The decision not only discouraged the Parishad, but also disappointed farmers across the state. Therefore, they promised to start a struggle against the government policy of giving water to the industries. On 26 October 2006, they protested against the government policy, organized under the banner of Orissa Krushak Sangathan and especially of the Paschim Orissa Krushak Sammanya Samity, led by Lingaraj and Ashok Pradhan. The leaders addressed the disgruntled farmers and encouraged them to continue the struggle. About 20,000 farmers participated in this movement (Tripathy 2006: 20–21; Samadrusti Televisions 2007).

In spite of such a strong protest, the Government of Orissa did not pay any attention to the problems and defied the demands of farmers. It prompted them to again stage their protest on 10 January 2007 and warned the government that they would continue their struggle till their demand was fulfilled. This time 40,000 people participated. They argued that the Hirakud dam was constructed by the sacrifice of more than 1,00,000 displaced farmers. In fact the displacement has been so widespread and compensation system so faulty that till date many farmers have not got it. They asserted that the farmers had not sacrificed their lives, homes, lands and livelihoods for the benefits of industries. Improvement of agriculture, especially of western Orissa, the most drought prone area, was one of the objectives of dam. The condition of the farmers in western Orissa was gradually improving by the introduction of irrigation in the region. But, the declining areas of irrigation in the recent years due to government's policy of giving water to industries has set process back again and the farmers in western Orissa are clamouring for water once again (Biswal and Rabisankar 2007: 17–18; Samadrusti Televisions 2007).

Response of the Government

The matters got even worse when, responding to farmers protests and movements, the state government callously and brutally ordered a police lathi-charge on them. More than 40 farmers got injured and many other journalists also received injury.[8] Such action by the government on farmers was not only condemned by the Krushak Samanaya Samity, but also by the entire state. The Samity expressed its anger to the state government and gave it a deadline of 72 hours to solve the issue (Sambada 2007: 5). On the next day, the Chief Minister of Orissa discussed the matter with ruling party members in detail and announced to increase the water level capacity of the Hirakud reservoir in order to give water both to the industries and to farmers (Sambada 2007a, 1, 8). The government gave three alternative proposals to increase the water level capacity of the reservoir. They were: extracting the deposited silt of the reservoir; increasing the height of the dam; dig the high places of the reservoir and make the reservoir deep. But these three proposals given by the government were vehemently criticized by the people as well as by many engineers. A retired Engineer of Hirakud dam, Karunakar Supakar, argued that earlier the government had expressed in its Technical Expert Committee Report that there was enough water in the Hirakud dam to irrigate the whole command area and now the same government was planning to increase the water level capacity. It clearly demonstrated that the government was hiding the truth in its report. Again, he emphasized that the three proposals made by the government have not been put into practice by any government in the world. In 2002, an engineer had estimated that 22 crores truck of silt will have to be extracted from the Hirakud reservoir in order to reach the original water level capacity. In 2002, approximately 12 thousand crores would be required to do this. In 2007, however, the cost of this endeavour had skyrocketed to 25 thousand crores. He further argued that in reality it was not possible to extract such huge amount of silt from the reservoir. Even if it were to be possible there is no place to keep such huge amount of silt. The dam had been designed for the highest water level capacity of 630 feet and it was not possible to extend the height. If it happens, the dam would break. The third alternative of digging the high places (small hills) inside the reservoir was also faulty. Digging them would not enhance the water level capacity. Er. Supakar dismissed these three proposals as 'unjustified and utopian' and demanded that the Hirakud dam be protected from the industrial encroachments and irrigation[9] be given utmost priority (Samaj 2007: 3).

In spite of such vigorous criticisms, the Chief Minister of Orissa declared, on 26 November 2007, to spend ₹ 200 crores, from the funds provided by the World Bank, to renovate the canals of Hirakud command area in order to irrigate all the lands in tail end area. This is clearly an attempt to shift the blame for decreasing water for irrigation on the lack of renovation of the Hirakud canal system and hoodwink the people into believing this by showing the money in the name of canal renovation. The real problem very clearly here does not lie with the canals but with the industries (Sambada 2007b: 1).

Notes

1. I personally met engineer Janaki Dutta of the Hirakud power generating units on 14 November 2007 at 3.00 p.m. and discussed about the details about the power generation.
2. The Hirakud dam has been designed for the highest water level capacity of 630 feet. But it has acquired the land beyond 632 feet. The acquisition of extra two feet by the government was primarily for the waves of the reservoir. Therefore, it further acquired 42 villages.
3. Lands were classified into six kinds (*bahal, berna, barcha, mal, bari and at*).
4. The houses in the submerged were classified into the following seven classes: (1) *Pucca* house with brick or stone and lime and masonry terraced roof. (2) Semi *Pucca* house with brick and mud lime plastering with country tile roof. (3) *Dhaba* house with walls made up of sun-dries and burnt bricks with country tiled roof over timber or bamboo and mud plastered roof. (4) *Dhaba* house as above with thatched roof for second low floor. (5) *Kachcha* house with walls of sun-dried bricks and country tiled roof. (6) *Kachcha* house with thatched roof. (7) Thatched small houses of poor people.
5. The study was done by me during my field study (1 June 2008–5 July 2008) of resettled camp (Kadalipal).

6. The study was done during the same period of my field study (1 June 2008–5 July 2008) in the resettled village (Jammal).
7. Hirakhand Nagarik Parishad is a registered social organization which consists of largely retired officials having social bent of mind and a few social workers. It raises their voices from time to time to influence the public opinion giving high publicity in order to help the policy makers to go in the right path and avoid public outcry.
8. I was present there and consulted with many farmers about their problems. Besides, I also discussed with Lingaraj Pradhan, the president of Orissa Krushak Sangathan, Ashok Pradhan the president of Paschim Orissa Krushak Sangathan Samannya Samity, and P. Samantara the President of Lokshakti Abhiyan, Orissa.
9. I personally met Er. Karunakar Supakar and discussed the details about the problems of farmers and irrigation of Hirakud dam.

6

Idea of Multi-Purpose River Valley Development (MPRVD) and Hirakud Dam

Dams have been built for the purpose of flood management, irrigation, hydropower production, navigation and to facilitate recreation. The fundamental promise of large dams in the twentieth century was abundant water and energy. However, at present we live in a thirsty world and it is likely to get thirstier. More than a billion people lack access to safe drinking water and high-yield crops and new industrial technologies often require enormous amounts of water. On the other hand, the demand of energy seems limitless. The forecasts of global fresh water and energy supplies in the decades ahead are often gloomy, and therefore it would be premature to write the political obituary of large dams. However, it should not be noted at the outset that all large dams have been failures. Indeed, dams have contributed to many benefits, even if these have often been distributed unevenly. The WCD has noted that, there has been a high degree of variability in the performance of dams, with some being plagued by numerous problems and others being relatively problem free. Multi-purpose dams face a difficulty which single purpose dams do not (Klingensmith 2007: 10–11).

Few benefits of dams are free of cost, as I have noted earlier, large scale displacement of millions of people and their forced eviction raise number of questions. Even rehabilitation efforts are made for the displaced, vary widely. Dispossession and rehabilitation occur in the context of the social relationship of power. The oustees mainly are ethnically distinct from those who gain most benefits from dams. Aboriginal or 'tribal' populations have been especially affected by large dams. Both proponents and critics of dams would acknowledge that dams necessarily create changes in uses of and access to rivers. This inevitably means a transfer of power over rivers as resources: away from local control and customary law, and to bureaucracies and state regulation, away from small agricultural producers and often from users of common lands and to large producers and to owners of privatized property. In many cases, there is also a transfer of access and control away from rural users of water to urban users and retailers. These transfers of power have effects on economic contexts, ways of living, and cultural resources of individuals and communities. Construction of dams in many cases has forced people into new physical setting, which are alien social worlds, functioning on unfamiliar lines, and into market economies which they enter at a structural disadvantage, having little educational, cultural or financial capital (World Commission on Dams 2000: 102–104; Sharma 2003: 907–908; Bharati and Rao 1999: 1374–1375; Cernea and McDowell 2000: 3663–3666; Cernea 1995: 266–267; Klingensmith 2007: 10–11).

Dams too have severe environmental ramifications. They alter hydrology and, therefore, ecosystem gets affected, both in the upstream and downstream. These changes hold implications for humans as well. There are also more immediate environmental problems for non-displaced people by the creation of artificial lakes. These social and environmental costs must be added to view the problems that impair the generation of benefits. The WCD found that dams providing irrigation water typically have under-performed as compared to the anticipation and thereby recoup their costs. It too has been compromised by water logging and salinization of soils. Hydroelectric dams did somewhat better, but still on balance did not meet generation targets. Meanwhile, multi-purpose dams have generally fallen short of their production targets than single purpose dams. The under-performance of all these kinds of dams may be, as in case of India, is low level of storage of water in the reservoir than the normal storage capacity. Problem aggravates in case of multi-purpose dams, largely they provide water for number of competing uses. Finally, the generation of benefits is impaired by cost overruns and schedule delays. Almost three quartes of the dams surveyed by the WCD experienced higher than expected costs. In case of multi-purpose dams, it exceeded the budget than single purpose dams. Approximately, half were completed behind schedule. Budget overruns raise the ratio of cost to benefits, diminishing the value of investment in a project (Klingensmith 2007: 10–11; World Commission on Dams 2000: 14, 44, 50–53; Sanmuganathan 2000: 22–57; Gritzinger 1987: 14–15).

The social and environmental costs of dams and the reduction of their benefits worldwide have become more evident and, thus, dams have become increasingly controversial. The political controversy over dams in the 1990s was the genesis of the formation of the WCD. Anti-dam activists, development agencies, including the World Bank, and corporations involved in building of dams organized a workshop on dam held at Gland, Switzerland in April 1997, to set up a commission, making its representation from government representatives, dam builders, academics and environmental activists. The purpose of the commission was to answer basic questions about the overall performance of dams – projected and actual. It concluded that dams indeed face all problems, those were earlier mentioned. On the other hand, it too conferred important benefits of dams. At the same time, it declared that the cost of these benefits had often been too high, sometimes unnecessary. Attention was drawn to unequal distribution of benefits of dams in comparison to various alternatives and it called for expansive and inclusive processes of decision-making, in which representatives of all those involved (including the displaced) would be heard. It also asked for thorough review of projects and more attention to alternative means of providing water and energy (Klingensmith 2007: 14–17).

There have been varied reactions on the WCD report. The report did not receive much mainstream media coverage in the United States and the government had little apparent reaction, perhaps because most of the dam sites in the country have already been used up and the rate of dam decommissioning now exceeds the rate of new construction. The funding agency, the World Bank, have been guarded in their reactions. Some activists and scholars have suggested that the report represents a good start, but owing to the WCD's diverse membership, does not go far enough in repudiating dams. Some governments have sharply rejected the report. China withdrew its representative on the commission before the report was finalized. India's Ministry of Water Resources questioned both the competency and good faith of the WCD and its report. At the same time, many Indian social and natural scientists, engineers and activists (for and against dams) have made major contributions to the WCD's knowledge base. One of the most important submissions to the WCD was a case study entitled *Large Dams: India's Experience*. The significance of the study is that it draws on a comprehensive body of scholarship from both government officials and anti-dam activists. It has vividly explained the investment on large dams since Independence, spent significant portion of its budget on dams. India has built more than 4,000 dams and between 695 and 960 dams are under construction, which is 40–56 per cent of all dams under construction worldwide (ibid.: 17–18; World Commission on Dams 2000).

The report has reported that about 40 million were displaced in India and they were never resettled and properly rehabilitated. It has criticized vehemently for poor effort to resettle and rehabilitate those oustees and calls for adequate attention for assessment of social and environmental costs, which were absent before 1978. Finally, the report concludes that most big dam projects of the 1990s would not have been viable, if the cost of mitigating the most severe impacts of them had been computed. In short, *Large Dams: India's Experience* affirms that many of the general problems outlined above have plagued India's dams (Klingensmith 2007: 18–19).

Idea of MPRVD and its Proliferation

The concept of integrated river basin management has been inspired by the United States during mid-1930s. Hoover dam, on the Colorado River is a conspicuous example of large multi-purpose storage projects, which uses a single structure to store water for multiple purposes. There were numerous projects at which two or three purposes were served and these storage projects were mostly in western Europe and western United States. In the Ruhr-Westfalian area, there was a series of 17 smaller dams constructed to store water for industrial and municipal supply, to generate power, and to reduce floods. Mill dams had provided water for industrial processing as well as to turn water wheels, and in some instances, they stored water for municipal purposes. Thus, in the Unites States, the Bureau of Reclamation combined electric power and irrigation storage at the Minidoka dam on the Snake River in 1909 and the Roosevelt dam on the Salt River in 1911. However, this is not to say that there were no combinations of two or more purposes in water storage or that the concept of multiple use was not current in public thinking. The irrigation barrage constructed under British supervision in India as a measure of famine prevention in drought areas had provided for some use of water for domestic purposes. These irrigation channels were also used for navigation, where it seems suitable (ibid.).

However, the much popularity of MPRVD, took place in the United States with the establishment of Tennessee Valley Authority (TVA) in the year 1933 and the construction of series of dams on the Tennessee River and its tributaries. It was designed to provide for the unified development of all the resources of the Tennessee Valley, an area that included a population of approximately four million persons and parts of seven southern states – Alabama, Georgia, Kentucky, Mississippi, North Carolina, Tennessee and Virginia. It constructed unified system of 26 dams on the Tennessee and its tributaries. By the latter half of the 1930s, the TVA's experience and its successes was emulated by other countries across the world. TVA, as an organization, was attracted by diverse places of the world, such as Rio de Janeiro, Cape Town, Paris, Shanghai, New York city and Stockholm. In the subsequent years, engineers, agronomists, economists, regional planners and numbers of scientific and technical staffs associated with TVA were deployed and charged with the mission of replicating MPRVD in different river basins and TVA module became the icon of modernity. Fifty foreign nations sent more than 700 technicians, government officials, students, business men and professional men and women to the Tennessee Valley to study the methods being utilized in developing the resources of the 41,000 square mile watershed of the Tennessee River and its tributaries. The gospel of TVA holds more promise abroad. From the 1940s, there were plans for new 'TVAs' in India, China, Palestine, Peru, Iran, Columbia and several other countries. There were plans for river development in these lands and it invoked TVA as their inspiration and called for TVA personnel to help in implementing them. TVA not only justified wide ranging dam construction across the world, but also inspired to set up number of corporations; Damodar Valley Corporation (DVC) in India, Corporation del Valle de Cauca (CVC) in Colombia and Khuzistan Water and Power Authority (KWPA) in Iran (D'Souza 2006: 184; Clapp 1948: 235–236; Klingensmith 2007: 68).

TVA and India

The success of TVA and its campaign for river valley development created an atmosphere of public and bureaucratic enthusiasm for large dams in India by 1945. However, one should not believe that the idea of dam construction in India is a post-colonial idea. The idea has already been initiated by the colonial administration, which had expanded and rationalized its survey of several potential dam sites, and made preliminary moves to establish the DVC and to build the Bhakra dam in Punjab and the Hirakud dam in Orissa. Soon after the Independence, the post-colonial government in India quickly took on further projects, and made the building of big dams a central part of its strategy for transforming India, a commitment to which it adhered for decades. In the 1940s, works on DVC, Bhakra and Hirakud were undertaken and in the 1950s, the government began to work on major projects on Rihand River, Koyna, Tungabhadra, Chambal and Krishna. In the 1960s, it began the Pong dam on Beas and started preliminary work on river Narmada and many others. It had undertaken 418 large projects during the period 1951 to 1970, and it grew sharply thereafter. By 2000, Indian state had completed nearly 4,000 dams (Klingensmith 2007: 211–215).

However, it raises many questions in our mind that what were the pushing factors for the government by that time, which led to sudden implementation and proliferation of dam construction in the post-colonial phase. Dams are equated with temples and agency to mitigate underdevelopment; however, in reality it is not. It has been mentioned earlier that the fact finding of the WCD report and number of studies on dams in India show that there is a dismal failure of dams' performances, and also that it has brought severe adverse socio-economic implications. Underdevelopment is still persisting in India in general and beneficiary areas of dams in particular. The intended objective to expand irrigation and hydropower production for the welfare of the people is now in question. Around 30 per cent of the India's population doesn't have electricity today and the condition of agriculture is in bad shape. Poverty is rampant among farmers community, and hence, practice of committing suicides among them has become a common phenomena in India.

Thus, 'dams as agency of development' has been interrogated by several quarters. Daniel Klingensmith (2007) has told that the so-called dams, called as agency of development, have been connected with nationalism in India. The political dreams of certain political leaders and anxieties among some of the bureaucrats have impelled them to recognize rivers and their valley since 1945. The integrated river basin management in India was largely promoted by many scientists, engineers and statesman, inspired by the United States. The dam building in India was reflected by both the bureaucratic politics of development and also nationalistic self-fashioning through the careers of many dam bureaucrats, like Meghnad Saha, A.N. Khosla and Kanwar Sain. Interest to build dams in India too grew in government circles and among the western educated public and a number of institutions came into existence to construct projects. Let us analyse how India as a post-colonial state promoted the construction of dams and tried to bring nationalism in its post-colonial era. In the wake of terrible Bengal famine of 1943, the Government of India appointed a committee to examine the causes of floods made by Damodar River and its possible future measures of prevention. The committee in its final report told the government that the nearest parallel to the Damodar is in the Tennessee Valley, USA. Acting on its recommendations, India asked TVA to send an engineer to study the Damodar's possibilities, and hence David Lilienthal sent William Voorduin for this purpose. This is the beginning of the Damodar

Valley Corporation (DVC), one of the best regarded development projects in the post-colonial era and even after 60 years of Independence, the DVC continues to make much of its TVA link. It also makes a point of acknowledging the crucial role played by the Bengali physicist Meghnad Saha (1893-1956) in guiding the committee toward its recommendations, and publishing TVA as a development model in India. His campaign for 'Indian TVA' on the Damodar is one of the most important achievements in his public life (ibid.).

Meghnad Saha is remembered as something of a Bengali national hero, and also as an important contributor to twentieth century astrophysics for his theory of thermal ionization. Saha was born in a poor family of petty shopkeeper in a village near Dhaka (now in Bangladesh). By virtue of his remarkable ability, family sacrifice, and luck in getting scholarship, Saha went from his village to Dhaka College to Presidency College, University of Calcutta, and eventually to the university's newly established Science College. During his stay in Presidency College, he came into contact with several political and scientific leaders of twentieth century. Among them were Subash Chandra Bose, P.C. Mahalanobis, Rajendra Prasad and Satyendranath Bose, and with the last he collaborated in producing the first English translation of Einstein's work on relativity. He got considerable professional success, which includes appointments to universities in Allahabad and later Kolkata, election to the Royal Society at an early age, and consideration for the Nobel Prize in Physics. In 1930s, he increasingly moved away from research to devote his energy to the popularization of science and its application to political problems. *Science and Culture*, which he founded and edited for many years, provided him an outlet for opinion on subjects of all kinds. He played an important role in the founding and deliberations of the Congress's National Planning Committee in the late 1930s. In pursuance of the same goal of modernist science for public sphere, he contested and owns a seat in the first Indian Parliament in 1952 (ibid.).

Saha was talking of 'river training', the technological solution to the 'problems' posed by Indian rivers. Such interest and explanation of river training was inspired from his home village, in the Brahmaputra delta, which was subject to yearly floods lasting up to four months. People in that region first learn swimming before walking. He was very much inspired by his mentor, P.C. Ray, who was a leader in flood relief efforts in Bengal, and encouraged him to join him in activities outside the classroom. Saha joined the relief programme during the Damodar flood in the year 1913 and 1923, which gave him publicity in Ray's flood relief drive in northern Bengal. In his numerous articles and speeches from the 1920s into the 1940s, Saha stressed the importance of rivers for the prosperity of Bengal and India, and explored the causes of flooding. The most influential publication was 'Catastrophic Floods in Bengal and How They Can be Avoided', published in the *Modern Review* in the year 1932. In the presidential address of the annual session of the newly established National Institute of Sciences in 1938, Saha insisted that a great civilization of the world all centred around their use of rivers, and advocated new opportunities for the revitalization of India through the systematic control of rivers based on scientific study. The disastrous Damodar flood of 1943 lent credibility to Saha's public statements on rivers and on the strength of his scientific prestige and his articles on flood control, Saha was appointed to the Damodar Flood Enquiry Committee and DVC, and the Indian TVA was recommended (ibid.).

A.N. Khosla was born in Punjab, educated at DAV College, Lahore, and then at the Thomson College of Engineering at Roorkee University at United Provinces. Afterwards he joined the irrigation engineering service around in 1920, under the British Raj. Throughout the nineteenth and early twentieth centuries, the British Raj had a leading role in the development of modern water management techniques and an increasing number of Indians joined the engineering service to transform India's infrastructure – building roads, railways, bridges, etc.

Among all the states in British India, Punjab was leading in the numbers of irrigation canals, weirs and dams. The Indian irrigation engineering establishment was pioneered by both British and Indian engineers. Sir Aurthur Cotton, known as Father of Modern Irrigation System, and Sir Ganga Ram promoted variety of projects in Punjab. On the other hand, Sir M. Visvesvarayya designed the Krishnaraja Sagar dam on the River Kaveri for Mysore, started in 1920 (ibid.).

Khosla was conferred the title of Rai Bahadur by the British Raj for his exceptional work in the Punjab government's engineering service, especially for a variety of irrigation and flood control projects. Afterwards he became the Vice-President of the International Commission on Large Dams, a dam construction advocacy group. The government decided that it was a worthwhile investment to send him to the United States with his junior colleague Kanwar Sain to study the design of high dams. Khosla's career illustrates both the exuberance and something of the bureaucratic empire-building that went into building India's new dams. He is reputed to have boasted from time to time that India would build the biggest dams on earth, projects outstripping anything undertaken by the United States or the Soviet Union. He was the chief government backer of the massive Hirakud dam. He firmly believed in creating new India by building large dams, acting himself as powerful state agency and at the same time building his own highly distinguished career. A.N. Khosla became the founder and first chairman of the Central Water and Power Commission and also a founder and first chairman of the International Commission on Irrigation and Drainage (ibid.).

The interest of building integrated river basin management in India was not only in the minds of few bureaucrats, but also among several political leaders, and especially of the first Prime Minister of India, Jawaharlal Nehru. His commitment to modern technology and scientific temper is very well known to the nation. During 1940s and 1950s, he advocated for promotion of modern science and drew himself to river valley development, which could contribute towards building the 'New India'. He was the most famous and most important celebrant of India's dams and cheerleader for India's engineers. He said in his speech, while inaugurating the Bhakra dam that, 'dams are modern secular temples of India'. In his first decade in office, Nehru took many opportunities to demonstrate moral or practical support for India's dam building efforts. In his first visit to the Unites States in October 1949, in spite of his busy schedule, he made a pilgrimage to Knoxvile to see the Norris dam (East Tennessee), which is indicative, America as the 'teacher' of development technique and India as its 'pupil'. He also sought out David Lilenthal in New York for advice on the possibility of a TVA employee for the post of chief engineer for DVC, a task normally not undertaken by a head of government. Nehru's enthusiasm for spectacular technology and his impatience of delay were no doubt useful for senior engineers like Khosla and Sain who wanted their projects approved quickly, and who sought to maintain service morale and public support. Even after his death in 1964, his affection for projects like Bhakra and DVC continued to be both an inspiration, a point of reference for the civil engineering establishment (ibid.: 255–265).

TVA and Hirakud Dam

It has been earlier described that the global model for MPRVD was pioneered in the United States in the early 1930s with the establishment of TVA, and it became the icon of modernity. The era was marked by the rapid decline of the European colonial powers and the emergence of the United States of America as the new hub of capitalist dynamism. Many governments adopted MPRVD immediately during their post-colonial phase with the help of American experts for its construction to speed up their economic development. D'Souza, in his study (2003), says that MPRVD scheme was inextricably linked to the political economy of decolonization and nationalism (D'Souza 2003: 82–83).

During the British Raj, in the year 1858, the flood problem in Orissa engaged considerable attention, when Sir Arthur Cotton, one of the pioneers of modern irrigation in India was asked to write a report on the harnessing of the waters of Orissa Rivers. After his inspection, he recommended the construction of weirs across Mahanadi, Brahmani and Baitarani, but these recommendations did not help the situation. In November 1927, the governments of Bihar and Orissa appointed a Committee of three experienced engineers to investigate the flood problem in Orissa. The Committee travelled extensively across the delta region and recommended that construction of flood control reservoirs was not feasible on any of the rivers. However, it suggested the opening up of estuaries in order to provide a free passage of floods into the sea and accelerate irrigation by pumping. In brief, the Committee did not recommend any kind of embankments, because it obstructed the natural flow of water (Government of India 1947: 7–8).

During the monsoon of 1937, there were high floods in deltaic tracts of the province causing considerable damage. Thus, the government sought the advice of Sir M. Visvesvaraya, who after a thorough study of old records made recommendations for framing a comprehensive plan to deal with the flood problems. He particularly emphasized on the feasibility of constructing flood control reservoirs on any of the three main rivers, particularly on Mahanadi. The Committee cited a similar situation of rivers like Mississippi and Ohio in United States of America being considered for the construction of flood control reservoirs on that basis of which he recommended flood control reservoirs on Mahanadi as well. Sir M. Visvesvaraya's Report of 1937 rejected the proposals of the earlier Committee of 1927. He gave two reasons for the rejection of earlier Committee's recommendation of constructing flood control reservoirs. The first reason was that the river was beyond the jurisdiction of the Government of Orissa, as it was part of Bihar province. Secondly, the huge cost of construction of reservoir on a big river like Mahanadi dismissed the idea of making any reservoirs. As Orissa got separated from Bihar province in 1935 and became an independent state, Sir M. Vivesvaraya's Report of 1937 to the flood problem of Orissa was most realistic and constructive. He not only recommended a solution to the flood problem of Orissa but also proposed the construction of multi-purpose reservoir which would solve the problem of drought and backwardness of the people of Orissa by promoting irrigation and generating electricity (Government of India 1947: 8–9; Sovani and Rath 1960: 1–2).

The Government of Orissa accordingly appointed a Committee in 1938 consisting of M.G. Rangaiya, C.C. Inglis and the Chief Engineer of Orissa to investigate the flood problem of Orissa. The Committee raised the question of the location of reservoirs. It recommended control of the discharges and the silt charges at the bifurcations of the rivers and emphasized on providing better outlets to the sea, strengthening the embankments and preventing breaches both in the embankments and in the natural banks (Government of India, 1947: 9–10).

Finally, in 1945, the problem was referred to the Government of India which passed it to the Chairman of the Central Waterways, Irrigation and Navigation Commission for investigation. The Chairman (Dr. A.N. Khosla) visited Orissa in May of that year. After his local inspection, study of previous reports and discussions with B.K. Gokhale, Advisor to the Governor of Orissa, and Rai Bahadur Brij Narayan, Chief Engineer, Public Work Department, he came to the conclusion that the only cure for the frequent floods in Orissa can be checked by means of storage dams. This was, in effect, the conclusion of Sir M. Visvesvaraya in 1937, revived in concrete form in the light of successful experiments in Multi-Purpose River Valley Developments (MPRVD) e.g, the Boulder Dam Project, the TVA, the Columbia Basin Development and the Central Valley Development in the US. Finally, in the year 1947, A.N. Khosla submitted a scheme for the 'Unified Basin-Wide Plan' of the Mahanadi valley to put a stop to colossal waste of the natural resources of Orissa. The scheme was designed to control, conserve and utilize the water of Mahanadi River primarily for flood control. Its auxiliary functions included irrigation, hydropower generation and navigation, with a view to raise the standards of living of the common populace in the area by banishing famine, eradicating malnutrition, controlling diseases and extending amenities of modern life to them (Government of India, 1947: 10; Mahalik, 2005: 74–76; Pattanaika Das and Mishra Undated: 48).

One could argue that the decision to construct a dam on the Mahanadi River was also a part of the broader calculus of post-colonial rule in form of increased centralization, and was not simply a case of seduction by the technologies of MPRVD. Although the colonial administration was very much enthusiastic to adopt MPRVD, its realization was challenged by the question of provincial autonomy and inter-provincial coordination. All work related to water management and allied activities was under the purview of provincial control. The Government of India Act of 1935 virtually debarred the central government from assuming executive authority over provincial subjects or from intervening inter-provincial matters. Under Section 135 of the Act, the Inter-Provincial Council could investigate or discuss matters related to provinces in a purely advisory capacity with no executive powers. But in post-war development, limitations were imposed on the provincial autonomy. It involved strong central direction, control and coordination in a field which was entirely provincial. In 1944, the Government of India asserted control over subjects listed as provincial concerns. On 9 September 1944, for example, Dr. B.R. Ambedkar, the Minister for Labour in the Government of India approved the establishment of Central Irrigation and Water Advisory Board (CI & WAB) and opposed any provincial consultation on the subject, which would cause delay. In December 1944, another federal agency named Central Waterways Irrigation and Navigation Commission (CWINC) headquartered at Delhi was created. The major concerns of this agency included irrigation, river control, conservation and control of flood waters, hydropower development, soil conservation, tidal problems and navigation. In response, almost all the provinces and princely states welcomed the formation of the Commission (D'Souza, 2003: 81–105).

Khosla's plan for the development of River Mahanadi in Orissa, in which he had an abiding interest, effectively illustrate his drive to expand his own and his agency's responsibilities and his commitment to gigantism. Although the flood enquiry committees of 1858, 1927 and 1938 made during the British rule rejected the idea of making any kind of dams or reservoirs on the River Mahanadi to check floods in Orissa, it was for the first time, in the year 1935, Sir M. Vivesvaraya suggested making of reservoirs in his report and finally it was approved by Khosla in 1945, and after the completion of the project, it was inaugurated by Jawaharlal Nehru in the year 1957 (Government of India 1947: 7–9; Sovani and Rath 1960: 1–2).

However, from the very beginning, Hirakud ran into problems. It was an enormous task to build such big dam, which is 16 miles long, the longest in the world. Three miles of the dam had to be built across the river bed and another 13 miles of dykes forming a wall running between the hills on either side of the valley. Such a dam required huge quantities of concrete, cement, masonry and earth fill. It too demanded great deal of machinery and CWINC decided that the dam should be built as much as possible by machine. Interestingly, some of the machinery bought by government was of the wrong short, unsuitable for job, creating serious bottleneck in the construction process. There were numerous petty scandals involved in the purchase of other materials. Works began two years before the designs were fully ready, resulting in false start and further cost overruns. In 1952, parliamentary investigation blamed Khosla for some of the improprieties and having overstretched himself, he had only been able to attend to Hirakud part-time, and could not adequately supervise his subordinates. Following the investigation, Khosla resigned, but by then he had created a considerable dam building establishment, and on the strength of the achievement he continued a public career of distinction and honour. In 1954, he was appointed as Vice-Chancellor of Roorkee University and then he became the member of Rajya Sabha in 1958 for two years. Afterwards, he became the member of Planning Commission in 1959 until 1962, and at last he was sent to Orissa as Governor (1962–68), the state in which Hirakud has been built. This last position was presumably given to him by virtue of his involvement with Hirakud, which was ultimately seen as a success (Klingensmith 2007: 218–220).

But the so-called success of Hirakud dam raises several questions when we empirically examine its objectives and actual performance. The failure of Hirakud dam could be easily seen through an empirical scrutiny of the fulfillment of its main objectives (see details in Chapter 4). The dam had four main objectives – flood management, hydropower production, irrigation and navigation, and in all four criteria, the performance has been dismal, to say the least. The claims about flood management fall flat when the empirical data shows that floods in river Mahanadi occur almost once in three years, causing havoc in Orissa (Government of India 1947: 50; Government of Orissa 2006: 132). The original plan envisaged a target of 1,524 million units of electricity generation. The target, however, even after 50 years, looks far-fetched as the dam generates only 948.65 million units of electricity generation as annual average which is merely 62.24 per cent of the original claims (Government of India 1947: 59–65; Government of Orissa 2006: 40). In terms of irrigation, the target was set at 10,94,953 acres, empirical records shows that only 6,11,583 acres of land – 55.58 per cent of the target – is annually irrigated (ibid.). The final objective – navigation, has only remained a distant dream (Government of India 1947: 67–70).

Further, the dam has submerged more lands and displaced more people than estimated in the feasibility report. It was estimated in the feasibility report that about 168 villages would be submerged covering 1,35,000 acres of land under the Hirakud dam reservoir. Out of which 70,000 acres will be cultivated land (Government of India 1947: 315–317). But after construction, it submerged 325 villages covering 1,83,000 acres of land. The submerged cultivated land was estimated at 1,23,000 acres. It submerged 291 villages in Orissa and 34 villages in Madhya Pradesh and displaced about 26,501 families (approximately 1,00,000 people) (Government of Orissa 2007: 1). The package of compensations to oustees was too less and many even didn't get a single penny. The government only rehabilitated 2,243 families, which is only 8.46 per cent of the total displaced people. Similarly, it replaced 8,468.80 acres of cultivable land, which is 14.52 per cent of the total submerged cultivable land (ibid.: 13–15). The displacement caused by the project forced the oustees to reel under various socio-economic crisis and marginalized them in various aspects of their life.

It has also led to large scale ecological and environmental changes in the region. It submerged 1,600 hectare of reserve forest and 20,400 hectare of village forests, heavily impacted on the pattern of rainfall in the locality. The construction of Hirakud dam brought a number of developmental activities, especially industrial and mining in the catchment area, which has been affecting significantly the water availability, its use, both in terms of quality and quantity, as well as inflows into and outflows from the dam. Hence, the intended objective and amount of hydropower production and irrigation need to achieve has been affected. The industrial effluents released from those industries into the Mahanadi too have been causing harmful effects on the people of these areas, their livestock, and cultivated crops. The increasing polluted water is no longer supporting the traditional occupation of agriculture and therefore forcing people of another means of livelihood.

There is not only a gap between the target and actual performance in terms of irrigation made by Hirakud dam, but also, there is uneven distribution of irrigation water leading to uneven development in the region. A study has shown that all lands are irrigated in the Head Reach Area (HRA), and on the other hand, only 35 per cent of the lands are irrigated in the Middle Reach Area (MRA) and only 18 per cent in the Tail End Area (TEA). The most marginalized among those areas are the farmers of TEA, who have been deprived of water in the *Rabi* season. Even, they don't get sufficient amount of water in the year of drought and therefore, uneven development has been persisting in the command area and now most of the farmers of the TEA are on the verge of poverty. Further, the scarcity of water has been compounded in the era of liberalization. The large numbers of industrial establishments around the dam periphery in the wake of economic reforms have caused the diversion of intended water for hydropower production and irrigation of the reservoir to industries, and hence, conflict has arisen between farmers and industries in the region.

Genesis of TVA

Making of any policy by any country is determined by their socio-economic and political situations. Therefore, it is very much essential to look into the genesis of the TVA in USA, the model which was replicated in most of the river basins of the world, and how far it is appropriate to apply in other river basins. The TVA was established in 1933 during the period of severe economic 'Great Depression' (1929–33). This was an attempt to address two discrete but inter-related developments that had come to haunt America in the 1930s: electrical modernization and fall-out of the great depression (D'Souza 2003: 83–84).

In the early decades of twentieth century, electric power had displaced steam power in the United States as the primary source of energy and it became the vital source of energy for industrialization. Technological advances in generation of power and transmission had enabled private utilities to create significant economies and, on the other hand, it too led to monopoly over electrical market. By 1926, private utilities controlled roughly 90 per cent of the entire electricity generated in US and, therefore, the monopoly of private utility ownership provoked a strong popular reaction against its pricing and service operations, which soon snowballed into a well articulated demand for federally controlled and regulated power production. Some of the public power advocates of the period such as Gifford Pinchot (Governor of Pennsylvania), George Norris (Senator, Nebraska), La Follete (Governor of Wisconsin) and J.D. Ross (Superintendent of Seattle City Lighting Department) argued that these private utilities were abstaining from their responsibility and keeping electricity rates artificially high. These critics demanded for the introduction of 'Giant Power'; the conservation of all primary energy resources into electricity and their common pooling into regional grids that would be coordinated to feed a nation-wide electrical transmission system. They also lobbied for the construction of federal power projects that would provide electricity with cheaper rates as compared to private companies (ibid.: 84–85).

Such reason forced the federal government to construct big dams and hydroelectric projects became pivotal to the federal involvement in electric production. The Grand Coulee, the Bonneville, Shasta and many other such large dams were built in this period by the federal government largely to dilute the monopoly control of electrical production by private utilities. A broad conception of river regulation and the idea of MPRVD, thus, received greater attentions by unifying the objective of flood control, irrigation, navigation, hydropower production, etc. However, the concrete realization of MPRVD was influenced by specific political and economic situation of the US. The execution of the TVA as a project followed from the ramifications of the Great Depression and the subsequent package of policies that the Roosevelt administration initiated to arrest the economic slide. The construction of dams on the river Tennessee was to revive the American economy by becoming an outlet for federal spending and generating employment (ibid.: 85).

The TVA was a part of the 'New Deal', attempts to solve America's economic and social problems in the 1930s. President Roosevelt set down the TVA Act as one of the priorities of his New Deal, and it was passed rapidly. The preamble to the legislation makes clear the main objectives envisaged at the time 'to improve the navigability and to provide for the flood control of the Tennessee river; to provide for reforestation and proper use of marginal lands in the Tennessee Valley; to provide for the agricultural and industrial development of the said valley; to provide for the national defense by the creation of a corporation for the operation of Government properties at and near Muscle Sholas in the State of Alabama, and for other purposes' (Bradshaw 1984: 209–210).

The TVA enjoyed broad public recognition not only in the United States, but in many of the nations allied with the United States in the Second World War. It was widely regarded as paradigm for 'world reconstruction', a model agency which showed how the liberal state could meet or avert the political and social challenges of economic crisis, revolution and war. It was a project of American liberal nationalism that promised to transform the impoverished, environmentally ravaged, war torn and 'backward' areas of the world, just as it had, supposedly, transformed the impoverished, environmentally ravaged and 'backward' American upper South. Following the TVA model, dam building programmes were promoted in Asia, Africa, Latin America and Soviet Union. The developmental policies of constructing dams, power stations, irrigation canals and so forth were linked to other things; such as American nationalism, liberal colonialism and postwar liberal modernism and it was followed in other countries too (Klingensmith 2007: 6–9).

However, looking at the present day theory and practice of river basin development, it has its limitation with regard to the degrees of development. A small proportion of flowing water of river systems of the world is being harnessed and applied for human good. With the exception of a few small drainage basins in arid regions, the water of no stream has been fully regulated or used. There are physical limits to such regulation and use, and it is related to conditions, which are partly technological, partly economic, partly political and partly ethical. Although the Nile and Tigris-Euphrates Valleys cradled the early civilizations of the eastern Mediterranean and still support their basic irrigation agriculture, and the flow of Rhine, Ohio, and Thames long have been essential to the industrialized populations along their banks, it would be inaccurate to regard any of them as having attained a high degree of development by contemporary standard of river regulation. Development implies two physical changes in stream flow. One is the regulation of flow by storage and diversion of water to cater to intended area during appropriate time, rather than as dictated by natural fluctuations over days, seasons and years. Under this definition, no stream can be considered fully developed unless, over long periods, its flow has been regulated. Only a few drainage basins are free from annual periods of water deficiency. In fact, dams engineering carry out such regulation in most, but do not resolute all problems in all situations. For example, the generation of hydropower depends on the volume of natural flow, fall and regulation. On the other hand, maximum potential output of hydropower and maximum possible use of water is influenced by the engineering feasibility of alternative designs, costs of constructions, alternative means of generating power, opportunities for marketing, availability of capital, investment alternatives, availability of labour force, accessibility of construction materials, and organization to plan and carry out the project. Likewise, in case of irrigation, it too depends on regulation and the consumptive use made by different assumed crops and cropping practices (White 1957: 157–184).

The actual accomplishment of Nile or the Tigris-Euphrates or the Rhine falls far short of full regulation or full use. Although the physical limits of development have been approached, the authorities concerned have not been reached. There are some exceptional cases too, for example, in spite of nearly complete regulation and use of flow, the upper Salt in Arizona and its storage structure in an area of heavy and chronic water deficiency halts all flood flows, and hence fails to yield enough and reliable supply of water to meet irrigation commitments. Even the Tennessee is not completely developed. Although, it has made its perfection in terms of reservoir regulation, it has failed to manage floods, when there are flood peaks. It too has failed to supply adequate water in the time of unusual droughts, which hinders economic life of people. There has been a gap between what is technically attainable and what really exists, may it be in Asia, Europe or in the United States. The comparative study of river systems shows that, the idea or the practice of river basin development can't be generalized, since, no way two rivers will be same. Each river is distinctive in characteristics of basin and flow. Very rare streams, regardless of size, are homogenous within their own drainage area (ibid.).

Contextualizing the case of Hirakud dam, question arises that how far was it appropriate to implement the TVA model of US on river Mahanadi. The political economy situation of the US and India was totally different during the post-war period and it was inappropriate to implement the TVA model in India. It was executed blindly without proper study and its socio-economic impacts. Although various studies on flood management possibility of River Mahanadi rejected the idea of MPRVD in the early years, those rejections were rejected in the name of nationalism in the post-Independence era. M.G. Rangaiya, an engineer of Mysore who opposed the construction of dam at Hirakud, in his report in 1946, opined that Naraj was the right place to construct the dam rather than at Hirakud. Flood management could be possible at Naraj by making a dam of 65 feet height. The dam at Hirakud would not check the flood if heavy rain occurred. Irrigation and generation of electricity could not be possible if there was scanty or medium rainfall. Today, after 50 years of construction, the dam is facing exact situations that M.G. Rangaiya had predicted (Supakar 2004: 15–18). But the government didn't pay any attention to his suggestion, and they implemented the TVA model of the US without understanding the technical feasibility of building dam at Hirakud on River Mahanadi and its socio-economic implications in the region.

7

Protests and Movements

Conventional political theory generally looks at protest and participation as distinct and contradictory forms of action, yet they can be complementary in a democracy. Protest movements may contribute to the legitimacy of democratic governance, as they significantly enhance the interactive space between the rulers and the ruled. As a mode of political participation, they put pressure on the state authority. At the same time, however, the type or the form of government in a country also affects the number and intensity of such protests. The structure and ethos of democratic regime, as against an authoritarian one, better addresses and responds to protests. Consequently, protest movements in a democracy are more extensive and less deadly than those under authoritarian rule (Swain 1997: 830–832).

Dam Constructions: Puzzling Trend

In the aftermath of the devastating Second World War, the western nations were compelled to refocus their energies on economic development. Similarly, the developing world, historically impoverished by its colonial masters and now being invigorated by decolonization process, saw for the first time the opportunity to chart the trajectory of its own economic development. National development strategies were framed which essentially followed the western ideas of modernization. 'Mega-projects', hence, became the dogma of the day. These countries realized that the availability of adequate infrastructure facilities was vital for the acceleration of their economic development. Dams, thus, became the symbol of development and their multi-purpose utility – generation of electricity, irrigation, flood control and navigation – contributed greatly to the growth of a nation (Joyce 1997: 1050–1055; Bandhopadhyay et al. 2002: 4108). However, it was only selected countries, which invested heavily in dam construction. In fact, the top five dam-building countries today account for nearly 80 per cent of all large dams worldwide. China, which had only 22 dams prior to 1949, has built around 22,000 large dams, close to half of world's total number. Other countries among the top-five dam building nations include the United States with over 6,390 large dams, India with 4,000 and Spain and Japan with between 1,000 and 1,200 large dams each. Estimates suggest that 1,700 large dams have been under construction in other parts of the world in the last five years. Of this, a total 40 per cent were reportedly being built in India (World Commission on Dams 2000: 8–10).

Post-Second World War, the rapid economic growth seen in subsequent two decades accelerated the global dam construction rate. At its peak, nearly 5,000 dams were built worldwide in the period from 1970–1975 (ibid.). However, the construction rate slowly declined coming down to 75 per cent in 1990s as compared to its peak period in 1970s, even when there was a growing scarcity of water and rising demands for hydropower across the world (Khagram 2004: 8–10).

Four sets of factors could be attributed to this puzzling trend – technical, financial, economic and political. Technically, the decreasing availability of sites contributed to the decline of big dam projects. Further shortage of funds financially choked some of these projects. For the Third World, oil crisis of 1973 and 1979 and the subsequent debt crisis caused huge problems, starving them economically. During this period, the international financial institutions (IFIs) pushed structural adjustment policies down the throat of many of these countries with a clear focus on privatization and public frugality. Moreover, other foreign leaders also shifted their strategies towards privatization, thus pushing these large and often profligate projects into backburner. Increased investments in other sources of energy-production like coal and gas, major advancements in renewable sources of energy and qualitative shift towards harnessing nuclear energy further decreased the rate of construction of dams (ibid.).

Dam Constructions and its Challenges

All these factors, however, only remain ancillary. The major hurdle that makes dams unviable is the mounting political opposition signified by ever-increasing protests against them. The enormous costs these projects entail in terms of environmental degradation and massive displacement make them vulnerable to rising demand of the anti-dam movements. By humbling the free flowing rivers, the dams have over the years invited strong reactions from various domestic environmental conservation groups across the world. In fact, some of the environmental movements that sprung up in 1950s and 1960s primarily emerged in response to dam building (ibid.).

With the dam construction being increasingly challenged in the First World by a vibrant environmental civil society network, dam construction declined. It forced transnational corporations and the funding agencies to shift their activities to developing countries. Therefore, approximately two-third of big dams built during the 1980s and three-quarter of them under construction during the 1990s were in the Third World. However, even the Third World did not remain impervious to protest movements. Spread of big dam movements, growing united struggles and campaigns waged by various environmental non-environmental organizations (NGOs), movements by indigenous peoples and growing human rights activism had already begun to affect dam projects from the 1970s (ibid.).

Growth of Transnational Actors

The empowerment of these actors contributed to the global spread and institutionalization of norms. Transnational actors over the years entrenched themselves strongly into the international systems and their empowerment has been the distinct feature of the twentieth century. From a mere strength of 176 in 1909, the transnational NGOs steadily grew in the first half of the century reaching 832 in 1951 and then literally exploded in the last half to reach a strength of 4,515 in 1988. The table shows that the transnational human rights organizations got doubled in the decade between 1973 and 1983, growing from 41 to 79 and shot up in the next decade to reach 190 in 1993. Transnational NGOs on environment as well as development issues also show a tremendous increase in the decade from1983 to1993. Significantly, these organizations, apart from numerous upsurges, also widened their networks and coalitions with their allied actors, both in domestic as well as transnational sphere, to promote social change in various issues. The increasing bonds between the First World and Third World NGOs have further empowered them and granted them legitimacy (ibid.: 11–12).

Table 1

The Growth of Transnational Non-governmental Advocacy Organizations

Issue Areas	*1953*	*1963*	*1973*	*1983*	*1993*
Human Rights	33	38	41	79	190
Environment	2	5	10	26	123
Development	3	2	7	13	47

Source: Khagram (2004:12).

Opposition to big dams has a long history every since their construction began in ancient times. But in recent years, this has come to command wider attention. In medieval England, boat owners opposed millers blocking rivers to create millponds to turn their waterwheels (World Commission on Dams 2000: 18–19).

However, it is after the 1950s, as dam building accelerated, that such protests have become widespread. Making their first footprints in 1950s, effective anti-dam campaign first started in the United States and many western European countries from the 1950s, thus contributing to the formation of transnational coalitions. The movement intensified over the decades resulting in a formation of strong network of anti-dam organizations. European NGOs like Ecologists, Survival International, Berne Declaration, Urgewald and European Rivers Network increasingly played a greater role in halting big dam projects funded by public authorities as well as private actors across the world. The most successful anti-dam campaign in the United States was the Grand Canyon campaign against the 525 feet-high Eco Park dam on the Green River. This dam was halted in 1963 after six years of constructions as civil society groups staunchly opposed it for economic and environmental concerns. The broader objective of anti-dam campaign was achieved with the establishment of the Environmental Policy Centre (EPC) by Brent Blackwater in 1972 which later became the Environmental Policy Institute (EPI). The EPI and American River Conservation Council (ARCC) established in 1973, jointly lobbied the federal government against large-scale water projects. The movement led by them with the support of others stopped more than 100 big dams in the US during the period 1972–83 (Khagram 2004: 177–183).

The campaign was further given a boost with the formation of another environmental organization – Environmental Defense Fund (EDP), which primarily emphasized on the negative impacts of large dam projects funded by the World Bank and other bilateral and multilateral agencies. Earlier, the campaigns were basically concerned with protecting river basins, preserving tropical forests and biodiversity. But gradually, anti-big dam protests were also made due to its large-scale involuntary displacement. From 1980s onwards, numerous advocacy organizations like Probe International in Canada, the European Network in France, the Association for International Water and Forest Studies in Norway, Friends of Earth in Japan, Both Ends in Netherlands, the Berne Declaration in Switzerland, Urgewald in Germany, Aid Watch in Australia, Christian Aid, Oxfam and Survival International in United Kingdom, International River Network (IRN), Cultural Survival and the Lawyers Committee for Human Rights in the US, the Narmada Bachao Andolan in India, the Movimento dos Antigidos Barregents in Brazil, the Assembly of Poor in Thailand and many others allied themselves together to reform and halt big dam building, around the world (ibid.: 183–189).

Apart from campaigns, the rise in number of publications against big dams by activists and researchers across the world further strengthened the movement. Edward Goldsmith and

Nicholas Hildyard published *The Social and Environmental Effects of Large Dams* in 1984 and its subsequent publications in 1986 and 1992, respectively. Their worldwide distribution expanded the transnational network. On the other hand, the California based bi-monthly *International Dam News Letter* started by Philip William in late 1985 and IRN publication of another newsletter *Bankcheck Quarterly* accelerated the movement increasingly institutionalized the transnational network (ibid.: 185–188, 198–199).

Further, several international conferences strengthened the transnational anti-dam network. The First International Conference of People by Dams, held in Curitiba, Brazil on 11–14 March 1997 and workshop promoted by the World Bank and World Conservation Union (IUCN), 'Large Dams, Learning from the Past, Looking to the Future', held in Ghana, Switzerland on 10–11 April 1997, attempted to bring together leading experts and representatives from various countries to initiate and open a transparent dialogue on the future of large dams around the world. The severe criticism by all participants and representatives of transnational anti-dam network resulted in an agreement to establish an independent World Commission on Dam (WCD) to review all big dams and their experiences built around the world to generate guidelines for the future (ibid.: 198–205).

On the other hand, numerous campaigns against World Bank-funded big dam projects during the 1980s and 1990s from many countries, and especially from developing countries, further strengthened and institutionalized the transnational anti-dam network. As a consequence, it compelled the World Bank to reduce its involvement in big dam projects and adopted various new policies on resettlement, environmental assessment, indigenous peoples and information disclosure (World Bank 1996: 75–108; Khagram 2004: 189–190). It influenced various other multilateral and bilateral agencies to adopt similar policies of World Bank – for instance the Inter-American Development Bank (IADB) in 1990 and Asian Development Bank is now in the process of consideration (Bartolome et al. 2000: 10). Individuals within the World Bank like Michael Cernea and Robert Goodland in their various reports began to explain about the socio-economic impacts caused by big dams during the 1970s and it brought certain reforms (Khagram 2004: 190–191; Michael Cernea 1988, 1995, 1998, 1999, 2000).

Environmental Movements and Protests Against Dams

There are ample historical evidences of dam-induced displacement in India since the British colonial period. The first struggle against dam-induced displacement took place in India in 1927 against the Mulshi Dam (hydroelectric project) located in western India, in the erstwhile Bombay Presidency (about 20 km. south-west of Poona). The struggle was a great landmark in the history of involuntary displacement in India. The project was owned by the corporate industrial house of the Tatas, and was commissioned in 1920. A total of 11,000 people were displaced and alienated from their traditional paddy field from this project. The indigenous people, Malva men, women and children staged a Satyagraha to stop the work on the dam site, but failed to do so. At last several hundred people were arrested and the struggle lasted only for two and a half years. It was largely due to the colonial British autocratic regime (Fernandes and Paranjpya 1997: 1–34).

In the post-Independence era, environmental movements and movement against dams too continued to progress towards defining a model of development to replace the current resource-intensive one that has created severe ecological instability. It has initiated a new political struggle for safeguarding the interests of the poor and the marginalized. Among the main environmental movements are Chipko Andolan, Save the Bhagirathi and Stop Tehri project in Uttar Pradesh; Save the Narmada (*Narmada Bachao Andolan*) in Madhya Pradesh and Gujarat; movements against mining in Gandhamardan Hills in Orissa; Appiko movement in Western Ghats; the campaign against the Silent Valley project in Kerala; opposition to Gumti dam in Tripura, etc. (Karan 1994: 32).

Chipko Movement: Soon after the Independence of India, Chipko Movement emerged as one of the strongest opposition to environmental degradation in the Himalayan region of Uttar Pradesh (now in Uttarakhand). It is relatively an inaccessible land of precipitous slopes, thin and fragile soils, and ample amount of water and forests, populated by subsistence farmers who derived a secure livelihood through their diligence and skills in a combination of terrace agriculture and animal husbandry. After the Indian-Chinese border conflict of 1962, an extensive network of roads was built throughout the region by the Government of India. The motive was clearly strategic; however, it caused severe environmental degradations wrought by road construction. The massive erosion of soil and landslides, blast shocks, degradation of forests and water sources, decimation of firewood, and massive social-economic dislocation of local populace forced them to protest against the government. The protests were further compounded by the construction of Tehri dam, on River Bhagirathi, which is the largest earth-filled dam in Asia, submerged thousand acres of land and forests and displaced about 40 thousand people. It was vehemently protested by local people and also by experts on the ground of local topography, seismic conditions and probability of high siltations. However, the protests and movements did not influence the government to halt the project (ibid.: 34–37).

When the region was connected with road infrastructure, tourism flourished to a great extent and the places of Badrinath, Kedarnath, Gangotri, Jumnotri and other places of Hindus became accessible to pilgrims and tourists in a day or two of travel by bus and taxi from plains rather than the weeks or months of trekking earlier. The new form of mass tourism taxed the capacity of the Himalayan environment heavily by populating the pilgrimage routes with hotels, restaurants, shops and other businesses. However, the beneficiaries of these kinds of development are almost exclusively outside entrepreneurs, their customers and political patrons. On the other hand, local people are not even employed in the enterprises brought to the region by the development. If employed, they are at most porters, milkmen, guides or manual labourers. The effects of the tourists on environment and local people received little or no attentions. The consumption of fuelwood alone is devastating because the demand for fires to cook food for the tourists and to warm their shelters far exceeds that of the local residents. Therefore, exploitation of the forest by outside entrepreneurs with government approval has been the source of conflict between Uttarakhand villages and the state Forest Department and Chipko movement emerges as one of the unique method to halt the government to cut trees. The word *Chipko* means to stick to or to hug to protect the trees from cutting. The movement's activists embrace the tree trunks to interpose their bodies between the trees and axemen. More than a dozen major and minor incidents of confrontations occurred during the 1970s. Each confrontation was non-violent and successful and the success of the movement led to increasing national and international publicity and recognition for the movement. Going from village to village, the Chipko activists prepared for each confrontation by informing people of the movement's purpose and inviting their participations (ibid.).

The movement has integrative effects at the national level by bringing together people from various regions of a diverse country and by providing a prototype of method and organization for similar problems elsewhere in India. In 1983, this method was followed to oppose reckless, illegal logging especially in Western Ghats of Karnataka. This movement was popularly known in Karnataka as *Appiko* movement. During the past century, there has been a progressive encroachment by the state on the rights and privileges of the people to forest resources. The people have resisted it in various parts of India, mainly through the Gandhian non-cooperative method of protest, well known as forest satyagraha (ibid.: 40–41).

Movement against Gandhamardan: The movement to save Gandhamardan hill of Orissa is one of the successful movements of 1980s. The mountain is located in both Bolangir and Bargarh districts of the state, where bauxite mining was proposed by Bharat Aluminum Company (BALCO) approved by the state. The mountain is not only rich in terms of forests but also it has its special mythical status, which attracts large number of Hindu devotees across the state. The place is famous as one from where Hanuman got the herbs for healing Laxman in the epic Ramayana. There are two prominent medieval temples stand by waterfalls on either side of the main ridge and the movement received lot of support from Hindu devotees (Padel and Das 2010: 89–95).

BALCO is a public sector company, started in 1965 in collaboration with Russian and Hungarian firms. It was found from their survey in 1972–78 that the Gandhamardan hill has the deposition of 200 million tonnes of bauxite, stretching 10 km. along the mountain ridge. The Chief Minister of Orissa, J.B. Patnaik, laid the foundation stone on 2 May 1983 at Paikmal in spite of strong local protests. The project was supposed to affect 104 villagers, and most of them belong to tribal community like Kond, Binjar and Majhi. The company built 9 km road up the mountain and set up a mining operations and an ore crusher at the top of the hill by December 1984. In the month of August 1985, it toppled the Garuda pillar at Narasinghanath, the medieval temple of Narasingha, the Lion Avatar of Vishnu (ibid.).

In response to it, a large number of devotees opposed to the project and the movement was further joined by large number of adivasis, dalits, women and activists across the state. Therefore the Ministry of Environment and Forests (MoEF), released a report of the project, which was very much critical especially of BALCO's claims that it would only affect 0.32 per cent of Gandhamardan's environment. The Ministry report stressed four major areas of impacts: biological diversity, water, *adivasis* and sacred sites. Responding to the report, Department of Mines commissioned a High Level Committee on project's environmental aspects, headed by B.D. Nagchaudhuri, (former Vice-Chancellor of Jawaharlal Nehru University, New Delhi). The Committee members visited the site twice during the period October 1986 to January 1987 and favoured their decision in support of BALCO. However, the decision was the beginning of the end and at last the public opinion forced the government to halt the project and the corporate arrogance came to an end and it led to people's victory (ibid.).

Save Bhagirathi and Stop Tehri Project: Protests against Tehri dam was one of the most influential movements during the 1990s. The Tehri experience has highlighted the crying need for scientific and socio-economic assessment of mega projects in the country. The dam is situated at the confluence of the Bhagirathi and Bhilganga rivers at the foothills of the Himalayan Terai, Garhwal district. It was commissioned in 1972, but the work began only in 1978. The height of the dam is 2160.5 m, said to be the highest in India. This rock-fill dam, expected to generate 3,500 mw of electricity, will irrigate 2,70,000 hectares of arid land between the Ganga and Yamuna. The dam submerged the small town of Tehri and 23 villages completely submerged in its vicinity and around 72 other villages along the river were partially submerged. It displaced about 70,000 people from their ancestral lands. The dam also flooded 1,000 hectares of cultivated land, 1,000 hectares of forest land and 2,000 hectares of pasture land (Centre for Science and Environment 1999: 145).

Resistance to dam started soon after the dam was commissioned. But the real anti-dam movement began to build up in the late 1980s. The Tehri Bandh Virodhi Sangharsh Samiti (TBVSS) started its protests against the dam. In October 1986, a writ petition was filed in the Supreme Court, on the ground that it would endanger life and property of the inhabitants of Munikireti, Rishikesh, Haridwar and places on the banks of Ganga. It was also opposed on the ground that the dam is located in a seismically active and tectonically unsuitable area. Thus, it would create insecurity among people staying on the rim slopes of the reservoir and it would cause large scale environmental destruction. Responding to the writ petition, a committee, headed by Sunil K. Roy, was set up in 1980 to assess the impact of the dam. However, the Roy Committee lamented that enough scientific data were not available for reaching definite conclusions on the geological and seismological aspects. The group, therefore, deputed the Wadia Institute, Dehradun, the National Geophysical Research Institute, Hyderabad and Roorkee University to conduct studies. The group was critical of the environmental degradation and large scale of displacement of people; there were differences of opinion within the working group. The Ministry of Environment and Forests (MEF) and Environment Appraisal Committee (EAC) also rejected the project on environmental grounds. However, in the light of an earthquake in 1991 at upper Bharirathi Valley, a five member of team (of earthquake experts) was appointed, which overruled the committee's recommendations. Professor Jai Krishna, an earthquake engineer of Roorkee University made a personal assessment in support of the project. As a result, MEF granted the project conditional clearance (ibid.: 145–116).

Sunderlal Bahuguna, famous environmental activist alleged that the commitments were not fulfilled by the government and thus he went on hunger strike in 1992. On the 45th day of his fast, Prime Minister P.V. Narasimha Rao promised to review the project and Kamal Nath, the Environmental Minister, too told in the Lok Sabha that the Centre was seriously considering withdrawal of environmental clearance to the project. However, construction work again started in 1995. Bahuguna again went on long fast in January 1995 and demanded the government to have an emergency review committee. But later, he was arrested and taken to the hospital. Again in the year 1997, he went on fast demanding the examination of safety, environmental and rehabilitation aspect of the dam. He asked the government to have proper rehabilitation of the displaced before the construction of the dam (ibid.: 146–149). After the hunger strikes failed to influence the government and most villagers were compensated with money, given land near Dehradun or house in the freshly built town of New Tehri, the movement started weakening. The anti-Tehri movement was not able to sustain itself even after it was termed the Himalaya Bachao Andolan (Save Himalaya movement) by its leader. The movement was naturally marginalized (Pathak 2005: 3637–3639).

Save Narmada: The most celebrated anti-dam protest in India is the mega Sardar Sarovar Project over Narmada River. The Narmada basin covers 94,500 square kilometres between the Vindhya and Satpura regions in central India. The Narmada valley is 1,300 kilometre long, contains large alluvial plains of Madhya Pradesh. To the west Narmada River, which is sacred to the Hindus, meanders through Gujarat, widening into a 25 kilometre of long estuary as it flows into the Gulf of Cambay. More than 20 million people live in the valley, mostly in villages. Among them, majority belong to tribal community (Bhils and Gonds), who occupy the forested uplands (Karan 1994: 37–39).

The Narmada valley is the site of one of the world's largest multi-purpose water projects: the Narmada River Development Project, which involves the construction of 30 large dams and many small ones on the river and its 51 main tributaries. It was proposed to transform the valley and the lives of its residents and will increase food production and hydropower production in Gujarat, Madhya Pradesh and Maharashtra. However, there was no detailed assessment of the overall environmental, technological and social effects of the Narmada project. It was estimated that these dams and reservoirs will displace about one million people and will submerge 3,50,000 hectares of forests land and 2,00,000 hectares of agricultural land. Major opposition was made against Sardar Sarovar dam by tribal groups, supposed to submerge almost 40,000 hectares of land and 250 villages. Financial assistance for this massive project was made by the World Bank, having approved the loans in 1985 before environmental impacts studies were completed. The bank was committed to environmentally sound development and issued guidelines for resettlement and rehabilitation of persons displaced by dams. It stated that no affected person should be made worse off by bank supported projects and it continued to fund the project. Despite its commitments, the Indian and the state governments could not meet the resettlement and rehabilitation guidelines and the social and environmental issues went unaddressed. Therefore, Narmada Movement began in the 1980s as a struggle for just resettlement and of people being displaced by the Sardar Sarovar dam (ibid.).

A number of protests were organized under the charismatic leadership of activist Medha Patkar in 1988. The movement is known as *Narmada Bachao Andolan* (Save the Narmada River) supported by the local people. The strategy of resistance was borrowed from the Gandhian Satyagraha including non-cooperation and civil disobedience, refusal to cooperate with project authorities, blocking all project-related works and refusal to leave their villages. Further the movement was strengthened by extensive studies on social and environmental impact of big dams. Activists and intellectuals from India and other parts of the world expressed solidarity with the struggle. The Narmada Bachao Andolan was confined not only against the dams constructed over the river Narmada, its influence also spread to other parts of India. It led to the withdrawal of the

Rathong Chu project in Sikkim in 1997 and the Bedthi project in 1998. Determined protests have led to the review of the rehabilitation package for Tehri and Koel Karo projects. By ensuring these voices, these movements have succeeded in compelling governments, both at the central and state levels, and powerful funding agencies like the World Bank to rethink their policies on displacement and rehabilitation (Centre for Science and Environment 1999: 136–141; Hemadri 1999: xxvi–xxvii).

Silent Valley Movement: In India, the Silent Valley movement in Kerala in the 1980s was the first significant victory for the anti-dam movement. The Silent Valley, one of the few remaining undisturbed rain forest areas in India, lie in the Malabar region, the least developed section of the state Kerala, at the southern end of the Western Ghats. This region is sanctuary of rare species of plants, ferns and endangered fauna that survive in the valley. During the early 1960s the state government began planning a dam for the Kuntipuzha River, which flows through the valley, to generate hydroelectricity as the basis for regional development. The projects offer a classic example of the dilemma between environment and development. Thus, opposition was made in the state led by Kerala People's Science Movement, a network of rural school teachers and local citizens, who brought the sharp focus of ecological consequences, specifically the possibility of extinction of species that had evolved over millions of years (ibid.: 39–40). Unlike other dams, a large number of people would have been displaced from their ancestral land and from source of their livelihood. The opposition came mainly from environmentalists apprehending the prospective destruction of the country's undisturbed areas of rainforest. The active support and involvement of international organizations like the World Wildlife Fund (WWF) and the International Union for Conversation of Nature (IUCN) along with the grand support from the local people ensured international attention and exerted heavy pressure on the Indian government to stop the construction of the Silent Valley project. As a result, the Government of India ordered the project to stop in 1983 (Centre for Science and Environment 1999: 134). The success of Silent Valley prompted the morale of dam opposition in other parts of India. In this project, environmental consequences were the central focus rather than displacement. However, alliances between environmentalists, scientists and tribal rights activists not only succeeded in securing the withdrawal of the Silent Valley but also the Bhopalpattnam, Ichhampalli and Bodhghat projects proposed over the Godavari and Indravati rivers (Hemadri 1999: xxvi).

Koel Karo Project: The Koel Karo project is the largest hydro scheme in the tribal region of the Ranchi district, Bihar. It was first conceived in 1954 to promote lift irrigation, rural electrification and industrialization in Ranchi, Gumla and Singhbum districts of Bihar and also to solve the growing power needs of Chotanagpur region. However, the Planning Commission cleared the project in 1973 and allocated ₹ 175 crore. The dam would submerge about 10,522 hectares of cultivable land, 5,666 hectares of barren land and 364 hectare of government land. It would displace 3,282 families in 26 villages of Ranchi district, of which 90 per cent are tribal (adivasis). The dam would also submerge over 15 *sarnas* (places of worship); the place was of immense cultural and religious significance for the nature worshipping *adivasis* (Centre for Science and Environment 1999: 152–153).

The project was not accepted by the local tribal, who demanded comprehensive rehabilitation – social, cultural and economic. The Koel Karo Jan Sanghatan (KKJS), a tribal organization vehemently opposed the project to let the project be implemented. In the early 1980s, the villagers were served with land acquisition notices. However, the compensations offered were too low to receive; therefore villagers moved court to stall the process. Though the government won the case, the Supreme Court formulated an improved relief and rehabilitation package for the people to be displaced and ruled that the people affected should be involved in all decisions that are likely to affect their lives (ibid.: 153).

Contradictions also arose over the beneficiaries of the project. Government claims that the project will provide water for irrigation, industrial, urban use and manage floods. On the other hand, activists suspect that the main objective of the project is to provide water to the growing industries in Jamshedpur and surrounding industrial areas and they continued the agitation. The movement was strengthened when it was supported by the Jharkhand Mukti Morcha (JMM) and the All Jharkhand Students Union. The sustained agitation of KKJS prevented the administration from laying the foundation stone of the project on 5 July 1995, when both Prime Minister P.V. Narasimha Rao and Chief Minister of Bihar, Lalu Prasad Yadav, had to cancel their proposed visit. The KJSS refused to discuss rehabilitation plans with the project officials and demanded to completely stop the project. Finally, the government closed down the Koel Karo project in 1997 (ibid.: 153, 156).

Bodhghat Project: Bodhghat dam was a part of the three-dam system to have come up on rivers Indravati and Godavari, the other two being Bhopalpatnam and Inchampalli. The Bodhghat in the Bastar district of Madhya Pradesh was conceived in 1972 and it was approved by the Planning Commission in 1979 with due clearance from the Central Electricity Authority and the Ministry of Environment. However, the project was stiffly resisted on the ground of massive destruction of environment. The region is moist peninsular and mixed deciduous forests, which is much richer than the Silent Valley of Kerala. Bastar is one of the last preserved forest areas of India; 66 per cent of the district is under forest cover. This area is also preserved habitat of the wild Indian buffalo, which is already declared an endangered species (ibid.: 156–157).

Therefore, environmentalists across the country strongly advocated preservation of the ecology of this region. Anil Agarwal, Director of Centre for Science and Environment, New Delhi, Martand Singh and N.D. Jayal of the Indian National Trust for Art and Cultural Heritage (INTACH), Kamala Chowdhary, former Chairman of the National Wasteland Development Board, environmentalist Sunderlal Bahuguna, Ashis Kothari of Kslapavriksh, New Delhi, and Bittu Sehgal of *Sanctuary* magazine, appealed to the Prime Minister to reject the proposal of the Madhya Pradesh government to stop the project. At the local level, the Bodhghat Virodhi Sangharsh Samiti lobbied against the project supported by tribals. At the international level, Survival International, a London-based leading interest coalition, maintained a concerted pressure against the World Bank and its decision to finance the project. They also put pressure on the Government of India to stop the project and thus the project was scrapped in 1994, even though the Government of Madhya Pradesh already carried out numerous structural activities in the project area (ibid.).

Bhopalpatnam Project: Bhopalpatnam was to be built on the river Indravati in Maharashtra. However, the project was stiffly resisted by the local people. The project would submerge 40,000 hectares of deciduous forests with teak and with other magnificent trees and displace about 99,000 people. Among them, about half were tribals. The project lies within the traditional home of one of India's ancient people – the Gonds. Critics argue that the real costs of the loss of these forests have been grossly underestimated by constructing such hydro-power projects. It is alleged that the real value of the standing forest would turn out to be 300 times more than what the government claims. On the Bastar side of Bhopalpatnam lays the Kutru National Park. It is one of the last remaining habitats of the wild Indian buffalo. About 8,094 hectare of this park was threatened with inundation. In 1983, Amte wrote a letter to the Prime Minister, Indira Gandhi, to stop the project. But the Prime Minister replied that the project must be continued for the sake of national interest. Having received such kind of replies, the local people protested against the project, led by Amte, tribal leader Lalsham Singh, Maharaja of Madhya Pradesh and environmentalist Sunderlal Bahuguna. The movement was also supported by the local Member of Parliament, Shantaram Potdukhe of the Congress (I) and L Dudiwar, Chairperson of Taluk Panchyat Samiti, Sironcha. The state forest minister also opposed the project and the government finally took the decision to postpone the project (ibid.: 157, 159).

Bedthi River Project: Like many other projects, Bedthi river project too has received continuous mass struggle against it. The project was initiated by the Karnataka Power Corporation Ltd (KPCL) and it was submitted to the Central Electricity Authority in April 1979. The Planning Commission cleared the project after some suggestion for environmental protection. A total of 6,065 hectares of land including 4,216 hectares of thick forests was to be submerged and around 4,000 people were to be displaced. The main issues of the contentions were submersion of large area under forest cove; displacement of tribals in the region; and, that the area already had about eight dams and one more would add to the environmental degradation of the place and lead to earthquakes. Continued protests against the dam forced the government take final decision and in 1998, the state minister M.C. Nanaiah said that the state would go ahead with project only after discussing it with people who would be affected by it. As a result, consecutive small projects of 10 mw each have been permitted to be set up at Panasguli, Hosakambi, Kelasi, Bairrekoppa, Dugganbali and Subguli, which would mean minimal submergence of land and less displacement (ibid.: 159, 162).

People's struggle against big dams in the post-colonial period has become one of the most prominent phenomena on the socio-cultural and political picture. The anti-dam movement has got a boost by the active support from the diverse groups – the displaced, economists, social science experts, human right activists, national media and international media. Anti-dam struggles have created an atmosphere that is more receptive to issues like displacement and environmental impacts and it has empowered the displaced or mostly the tribals to raise their voices against their human rights Violations. Even if there were anti-dam struggles during the pre-Independence period, these were not organized. The resistance in the post-Independence has been more organized, sustained and has had a profound influence on the entire discourse of dams, displacement and development (ibid.: 133–134).

Anti-Dam Movements and its Impact

Dams were considered as the monuments of progress, which followed after Independence. However, in the later years, dams as development projects were resisted and protested for its large scale displacement of people and environmental degradations. The role of the anti-dam movement was instrumental in bringing about this change in the mind of people and government's attitude. At first, it learnt and made people conscious. It gave directions to the political rights of tribals and the downtrodden, who face the adverse consequences of the development projects. Activism in the Narmada valley is historically important, which mobilized people and made them politically active. It imparted environmental consciousness and people started demanding their rights, for control over natural resources and to retain their traditional lifestyle. People demanded information on all technical aspects of the dam, like the extent and schedule of submergence, basis for estimating the cost-benefit, and methods of assessing the value of land (ibid.: 162–163).

One of the most important lessons of the anti-dam movement is that it brought integrated, holistic, multidisciplinary planning from the beginning. Earlier, planning of dam construction was on the basis of engineering convenience, which was not suited to geo-climatic region. Environmental and other concerns were often disregarded, which were not the integral part of the project planning. The people, who will be affected, were not consulted and there was no sharing of information to them (ibid.).

The anti-dam movement forced the government to review environment and rehabilitation aspects of dam projects and it changed the government attitude towards addressing displacement and environmental degradation. It forced the government to bring transparency, setting up independent committees in consultation with the movement's leader to review the safety, rehabilitation and environmental aspects. Various committees were made to review the environmental and rehabilitation aspects of the Tehri dam project. Even, a high level committee was constituted to examine the safety aspects, but at the end, it was left to the bureaucrats to manipulate the government to clear the project. The anti-dam movement too informed and alerted government that rehabilitation work should precede over the construction work of the project. This was introduced when the Sardar Sarovar and Narmada Sagar projects were granted conditional clearance (ibid.: 164).

It manifested in a shift in policy and legislative measures. For example, the Land Acquisition Act of 1894, which provided only cash compensation for the oustees, was amended in 1984 to facilitate 'land for land' compensation policy. The resettlement and rehabilitation policy of the governments of Gujarat, Maharashtra and Madhya Pradesh, namely, the states affected by the Sardar Sarovar project, was an improvement on the award of the NWDT. There was a feeling of gaining ground in government circles that there is no future in alienating people by building high dams. Even the powerful engineers lobby, who have so far called the shots, were increasingly worried of pursuing such major projects because of doubtful long-term viability and value. One of the major steps as a result of anti-dam movement is the consideration for a National Rehabilitation Policy, which seeks to give a just rehabilitation to the oustees. The policy is being drafted by the Ministry of Rural Development (ibid.: 164–166).

One of the biggest impacts of anti-dam movement was on the role of funding agencies. World Bank is the largest funding agency to dam construction in India. Sardar Sarovar project was funded by the World Bank. Initially, it ignored the anti-dam movement. However, the local opposition and international pressure led to much rethinking and debate within the Bank. It set up an independent review team, submitted its report in 1992, which was very much critical of the project. In 1993, the World Bank funding was discontinued. This development had a long lasting effect on the future of dam construction activities not only in India, but in other developing countries as well (ibid.: 166).

Case of Hirakud Dam

There was large-scale opposition from the local population as soon as it was decided to build dam at Hirakud. Radhakrishna Biswas Rai, a minister, first announced that the construction of Hirakud dam would save the delta from the ravages of floods, the most fertile land and thickly populated region of Orissa. As the first notification came on 13 September 1946 for acquisition of land in 95 villages, strikes were organized in Sambalpur town. This provoked the local people to ask as to why Sambalpur should suffer for coastal people. It is irrational to displace a large number of people from Sambalpur district in order to save the prosperous coastal delta from floods. Besides, it was opposed by

M.G. Rangaiya, ex-chief engineer of Mysore. He subsequently released a 26-page report and submitted his logics for the opposition of the dam at Hirakud. The argument made by him was that the dam would destabilize Orissa's economy due to absence of any proper estimate of cost and benefit of dam construction. The idea of generating electricity from Hirakud was shown by him to be very expensive as there was no natural waterfall and he rather suggested for the creation of a thermal power plant. He views the idea of navigation as a doubtful proposition, especially in the context of modern locomotives and changes the nature of trade and commerce. He also expressed his doubts on the life span of dam due to the large amount of silt in the river (Baboo 1991: 2374; Pattanaik Das and Mishra undated: 53; Supakar 2004: 10–11).

The local leaders along with some bureaucrats took an active role in opposing the dam construction. Prominent among them were Bodhram Dube, L.N. Mishra, Prasanna Panda and Sradhakar Supakar. Dube pointed out that the flood problem in Orissa dates back to the introduction of canals at Cuttack and Puri; therefore construction of dam would not solve the problem. Further, he argued that the establishment of various industries in and around Sambalpur district would not provide any benefit to the local people. Rather the rich and the industrialists would reap the benefits and local people would be employed as daily labourers. On the other hand, L.N. Mishra added that it would result in the loss of fertile land of 310 villages and loss of abundance of minerals. Bharat Nayak, an ex-deputy minister, in a public meeting rejected the idea of dam construction and argued that the displacement would create enormous emotional problems for those leaving their ancestral homes. Alternatively, he suggested, due attention should be given to the implementation of embankments (ibid.).

In the last week of May 1947, a satyagraha was started under the leadership of Danardan Pujari and his wife Kamala Devi in Sambalpur to stop the construction of Hirakud dam. But they were arrested on the first day of the satyagraha. Later, on the third day, Sradhakar Supakar, Prasanna Kumar Panda, Satrughna Panda, Chintamani Hoto, Purander Panda and

Srinibas Mishra, etc. were arrested. Sankar Prasad Mishra, Hari Charan Padhi, Durga Prasad Mishra, Sairendri Nayak, Anantaram Mishra and Ishwar Prasad Mishra, etc. had made immense contribution in this movement. A series of public meetings were also organized under the leadership of *Gountias* (village headman) of different villages. A resolution was adopted to oppose the project work through peaceful agitation under the leadership of Bodhram Dube, to take steps to separate Sambalpur from Orissa and to demand withdrawal of prohibitory laws. The demand of separation of Sambalpur from Orissa formed an important basis for the organization and mobilization of the people, the local Congress leaders withdrew from the agitation charging it for conspiracy by the former rulers of Garjats against democratic planning and development and trying to perpetuate feudal rules propagated against national sentiments. The move was also disapproved by the state and the Indian National Congress including Gandhiji, and gradually the movement lost its momentum. The arrest of important leaders, withdrawal of Congress leaders and betrayal of Bodhram Dube stopped the progress of the agitation. Finally it came to a halt once the construction work was inaugurated on 12 April 1948 by the Government of Orissa (ibid.).

Although the nature of the domestic opposition to the Hirakud dam was strikingly similar to domestic anti-dam struggle that emerged in the greater numbers in India during the 1970s, it failed to prosper due to the absence of NGOs, transnational allied advocacy networks, legitimized global norms on human rights, indigenous peoples and the environmental lobbies. The domestic resistance to Hirakud dam was politically too weak to influence the institution of India's democratic regime to either halt or even reform the Hirakud project (Khagram 2004: 36–37). Besides the failure to attract larger alliances, the resistance to Hirakud dam was overshadowed by the nationalistic rhetoric of nation building in these early years of post-Independence (Hemadri 1999: xxvi). The Prime Minister of India, Jawaharlal Nehru, had, in fact, once said, 'Dams are temples of modern India'.

The resistance to Hirakud dam could be justified on the ground that none of the four objectives – flood management, hydropower production, irrigation and navigation of the dam has been fulfilled even after its 50 years of completion. The empirical scrutiny of the Hirakud dam shows that, none of the four objectives are fulfilled. The empirical data shows that floods in river Mahanadi occur almost once in three years, which cause devastation in Orissa (Government of India 1947: 50; Government of Orissa 2006: 132). Estimation was made in the feasibility report that the dam would generate 1,524 million units of electricity, but it is only generating 948.65 million units as annual average, which is merely 62.24 per cent of the original claims (Government of India 1947: 59–65; Government of Orissa 2006: 40). In terms of irrigation, the target was made to irrigate 10,94,953 acres, but in reality, it is irrigating only 6,11,583 acres of land annually, fulfilling 55.58 per cent of the target (ibid.). Finally, the objective of navigation remained a distant dream (Government of India 1947: 67–71).

Further, it has shown that the Hirakud dam has submerged more lands and displaced more people than estimated in the feasibility report. Estimation was made in the report that about 168 villages would be submerged covering 1,35,000 acres of land and out of which, 70,000 acres will be cultivated land (Government of India 1947: 315–317). But after the construction of the project, it submerged 325 villages covering 1,83,000 acres of land, including the 1,23,000 acres of cultivable land. Out of the 325 villages submerged, 291 villages in Orissa and 34 villages in Madhya Pradesh were submerged and displaced about 26,501 families (approximately 1,00,000 people) (Government of Orissa 2007: 1).

Most of the displaced villagers were dependent on agriculture and for many centuries, who lived in close proximity of Mahanadi and its tributary, the Ib River. The rates of compensation for likely submerged lands proposed in the feasibility report ranged from ₹ 50 to ₹ 1,000 per acre, according to their classification in terms of productivity. It was classified into six kinds (*bahal, berna, barcha, mal, bari* and *aat*). But in actual, they were paid more or less in a uniform rate ranging from ₹ 200 to ₹ 600 per acre, which was much less than the market value. Similarly, the amount of compensation for submerged houses was too less to construct a new home elsewhere (Government of India 1947: 315–319). In the process of rehabilitation, the government resettled 2,243 families in 18 different rehabilitation camps, which is only 8.46 per cent of the total displaced people. Similarly, it replaced 8,468.80 acres of cultivable land, which is 14.52 per cent of the total submerged cultivable land (ibid.: 13–15). Lack of proper compensations and rehabilitation by the government forced the displaced people to move different places to settle themselves on their own initiatives. It resulted in severe livelihood crisis, health hazards and diseases made them victims in their initial period of self-resettlement. Submergence of their lands under the Hirakud reservoir forced them to reel under various socio-economic crisis and marginalized them in various aspects of their life.

In the feasibility report of 1946, M.G. Rangaiya, an engineer of Mysore, opposed the construction of dam at Hirakud and suggested that Naraj was the right place to construct the dam. He opined that flood management could be possible at Naraj by making a dam of 65 feet height. Flood management can't be possible at Hirakud, if there is heavy rain. In case of irrigation and hydropower production, the performance will also be too less, if there is scanty and medium rainfall. Now, after 50 years of completion, the dam is facing the exact situations what M.G. Rangaiya has predicted. But the government didn't pay any attention to his suggestion then (Supakar 2004: 15–18).

8

Conclusion

Referring to the development strategy after the Second World War and economic growth as its indicator, all the countries, both in devcloped and developing, started to grow their economy. The availability of adequate infrastructure facilities is vital for the acceleration of economic development of a country. Governments across the world have given high priority to investment in sectors such as railways, roads, power, telecommunications, ports and industries, etc. Thus, dams are the outcomes of this process and symbols of development and their multi-purpose utility – generation of electricity, irrigation, flood control and navigation – contributed greatly to the growth of a nation. There was rapid increase in large dam building during the last century. In 1900, there were only 600 big dams in existence and many of them were built in Asia and Africa. It grew about 5,000 big dams by 1949 and by the end of the twentieth century over 45,000 large dams were built over 140 countries. Thus, over 90 per cent of big dams were built over the last 40 years. United States, Spain, China, Japan and India are the top five dam building countries in the world.

Electricity generation is an important reason for building large dams in many countries either as the primary purpose or as additional functions for other purposes. Hydropower currently provides 19 per cent of total electricity supply over 150 countries. One-third of the countries in the world currently depend on hydropower for more than half of their electricity needs. Agriculture plays a greater role in the economic and social development and is seen as a vehicle for development, poverty alleviation, rural development and solution for food insecurity. It is estimated that dams irrigate 40 per cent of the total irrigated land worldwide, which contribute about 12–16 per cent of the world food production. About 13 per cent of all large dams in the world – in more than 75 countries – have flood management functions.

While many have benefited from the services dams provide, but their construction has led to many significant social and human impacts, particularly in terms of displacement. Thus, affected people face a broad range of impoverishment risks that includes landlessness, joblessness, homelessness, marginalization, food insecurity, increased morbidity, loss of common resources and social disarticulation that result in a loss of socio-cultural resilience. Compulsory displacement that occurs for development reasons, embody a perverse and intrinsic contradiction in the context of development. They raise major ethical questions, because they reflect an inequitable distribution of development's benefits and losses. Nevertheless, the involuntary displacements caused by such programmes create major impositions on some population segments. It restricts population rights by state-power intervention. This raises major issues of social justice and equity.

Large dams have significantly altered many of the world's river basins with involuntary displacement and its impacts on the 'livelihoods' and socio-economic foundation of millions of people. Environmental changes and social disruptions caused by large dams and associated infrastructure developments can have significant adverse impact on 'human health' for displaced communities. 'Gender' is the vital element, while understanding the impacts of involuntary displacement. In the process of involuntary displacement, conflicts also arise in many cases between the displaced people and 'host communities'. It put heavy pressure on the resources of host communities and competitions take place over the sources of livelihood and employment.

Government takes initiative to resettle and rehabilitate the displaced people. The main objective of the resettlement is based on collective negotiations with the affected people, making plans and implementing it as a development project to restore pre-project income and living standards. There are two strategies (land based strategy and non-land based strategy) generally in practice for the socio-economic re-establishment for the oustees. The former includes the adequate financial compensation for lost property and providing economic opportunities to re-establish their agricultural based livelihood sources, while the latter strategy is the diversification of the livelihood sources to other activities. However, implementation of such programmes by governments is not so encouraging and it has produced indignation and bitterness rather than mitigating of displaced grievances. In this context, Rawl argues that the project, which promote greater good for greater number impoverish 'oustees' and compel them to sacrifice for 'greater good' are intolerable. The scheme or system, which creates advantage for some parts of community, is unjust, unless it creates benefits for all.

At the time of Independence, India was in the stranglehold of stagnating per capita national income, static agriculture production, poorly developed industries, inadequate infrastructure, mass poverty and extreme unemployment etc. Among all the problems, food scarcity was worrisome to the whole nation. Indian agriculture was in deplorable condition. Its productivity had been declining and was generally among the lowest in the world. Despite the fact that nearly 70 per cent of the working population was engaged in cultivation; the country was not self-sufficient in food grains, and had come to depend on imports of food grains. Thus, huge and organized effort was made on a national scale to achieve substantial progress on the socio-economic front. Thus, planning was accepted as the key strategy of India's developmental efforts. An important part of such planning was tc raise the agricultural and industrial growth by utilizing country's great water resources, particularly through multi-purpose river valley projects. Construction of dams was given high priority in India's each Five Year Plans and over 15 per cent of the total plan outlay was invested on dams. Till now, irrigation and hydropower generation have been the major objectives of water resource development in India. Soon after India's independence, various multi-purpose river valley projects were constructed. Some of the important river valley projects are

Hirakud dam project in Orissa, Bhakra and Nangal project in Punjab, Tungabhadra River project in Tamil Nadu and Andhra Pradesh and Kosi River project in Bihar.

Hirakud dam is part of the Mahanadi Valley Development – the Unified Basin-Wide Plan. It comprises the construction of series of dams on River Mahanadi at Hirakud, Tikarpada and Naraj to control, conserve and utilize the water of Mahanadi for the primary purpose of flood control and auxiliary function of irrigation, hydropower generation and navigation, with a view to raise the standard of living of the common man by banishing famine, malnutrition, disease and to extend their necessities and amenities of modern life. The construction of Hirakud dam started in the year 1948 and completed in 1956.

With the construction of Hirakud dam, it not only managed floods to certain extent in the coastal Orissa, but also brought development in the region in terms of irrigation and hydropower generation. Development of farmers of the region has accelerated to a great extent with the opening of irrigation. Industries and mining activities flourished to a large extent with the supply of hydropower. Thus literally, the hydropower system became the fore-runner of the industrial growth, economic progress and overall development of the state. Apart from meeting industrial requirements for hydropower, it also supplied electricity to domestic, municipal and township needs catering to many parts of Orissa.

Although, the Hirakud dam brought economic progress of the state, it has not fulfilled its target as per the original plans envisaged in the feasibility report. It has failed to perform in every front. The primary objective of managing floods is totally failed. The empirical data shows that floods have been occurring once in every three years in the River Mahanadi. Similarly, there are underperformances in case of irrigation and hydropower production. The objective of navigation has only remained a distant dream. The dam has also submerged more land and displaced more people than the estimation in the feasibility report or the original plan.

A new department known as 'Hirakud Land Organization' (HLO) was set up by the state government of Orissa for orderly evacuation, resettlement and rehabilitation of the displaced people. However, the package of compensation, process or resettlement and rehabilitation is completely unsatisfactory. Compensations for the submerged land and houses made by the government are too less, which are much below the market value. Many of them also didn't get any compensation. Following the land based strategy of resettlement and rehabilitation; the displaced people were rehabilitated in 18 different camps, which resettled only 2,243 displaced families, which merely constitutes 8.46 per cent of the total displaced. The oustees, who were really wanted to be rehabilitated by the government were just dumped in those identified camps (full of jungles and wild animals), which were devoid of basic amenities of life and without facilitating basic sources of economy, roads, schools, hospitals, clubs, markets etc. It has been shown in my study that the government neither provided the oustees lands for cultivation in the resettlement camps nor the compensations for land reclamation activities. Although, the government had the plan of resettlement and rehabilitation programme for the oustees based on land-based strategy, mentioned in the feasibility report, there was no provision with regard to non-land based strategy nor the government has initiated such programmes later to raise the sources of livelihoods.

The most affected among them are the landless and tribals, primarily were dependent on forests resources for their livelihood. But the absence of their legal ownership over lands deprived them of compensations and marginalized them the most in the whole episode of displacement. They were forced to shift their source of livelihoods from the forest economy to market economy, fled to nearby villages or urban areas in search of employment. Besides tribals, the non-farming communities (barbers, car painters and blacksmiths), village priests and village guards were the most victims, who were maintaining their livelihood by practicing their occupation from generations. With the displacement of people, they lost their clients as well as their occupation and livelihoods. The same problems were faced by the petty businessmen as well. But, the nature of the problems of village priests and village guards faced are little bit different from the non-farming communities. Although, they had possessed quite sufficient amount of land, were deprived to get compensation from the government. Since, these lands were gifted land by the villagers to them for their services, and hence, the absence of legal ownership of such lands deprived them of compensations, and the livelihood crisis persisted among them.

In fact, the degree of livelihood crisis varies from the completely submerged villages and the partially submerged villages. In case of former, people permanently shifted themselves to new places according to their own suitability. But, in case of latter, many of the people still decided to stay in the periphery of the reservoir, which has brought new social changes among them, especially in terms of occupations. With the shift of accustomed environment and ecology, the displaced people faced severe 'health crisis'. Among all diseases, Malaria was very much rampant among them, primarily due to the region was densely covered with forests. The severity of socio-economic impacts caused by displacement was even more on 'women' than men. The study has shown that the oustees have reeled under severe 'impoverishment risk'.

Although, Hirakud dam brought massive economic progress of the state and has become the fore-runner of industrial growth and accelerated the area of irrigation and flood management etc, it is ironical that none of the oustees got any of the benefits of the project. Rather, they were deprived in the name of 'greater good of the greater numbers'. The oustees were not given even a piece of land in the irrigated area as compensations to maintain their livelihood. They were also not given electricity produced by the dam and most of it is supplied to industrial, mining and urban sectors. On the other hand, the unsuccessful resettlement and rehabilitation of the oustees created a new inequality by depriving them not only of 'income and wealth' but also of social goods in two other ways: 'liberty and opportunity' and 'social bases of respect'. Hence the project Hirakud which is creating advantage to some and depriving others is unjust.

Over half a century after the construction of the dam, its catchment, reservoir and command have undergone considerable economic, environmental and ecological changes. These changes have significantly affected water use and availability, both in terms of quality and quantity, as well as inflows into and outflows from the dam. The rainfall pattern in the region too has changed considerably and therefore impacted a lot to farmers. Looking at the above scenario, management have been changing its priorities of natural resources in these different spaces and therefore it has been putting strong impacts on the evolution and manifestation of the water conflicts around Hirakud.

In spite of its huge achievement of Hirakud dam, questions have been raised for its uneven development in the command area. Study made by Ahmedabad based 'Development Support System' says that there is uneven distribution of irrigation water leading to uneven development in the region. It states that all lands are irrigated in the Head Reach Area (HRA), and on the other, only 35 per cent of the lands are irrigated in the Middle Reach Area (MRA) and only 18 per cent in the Tail End Area (TEA). Further, a study made by Meher (2011) states that although in the initial years the dam was irrigating the whole command area, however since 1990s, the area of irrigation gradually started to decrease and have been affecting the livelihood of the marginalized populations. The small and marginal farmers comprising mostly of the tribals and other marginalized people living the tail end of the command area are now deprived of irrigation water. They are not in a position to grow any second crop during *Rabi* season (winter and summer crop) due to acute water scarcity. Also, during drought years they do not get adequate water to protect their standing paddy crop in the *Kharif* (rainy and autumn crop) season, which is a timing of pressing need.

The scarcity of water at the tail end area of canal is largely due to drawl of huge quantities of water by various industries from the reservoir for industrial use. Thus, conflicts have been going on between farmers and industries over the water of Hirakud dam. Soon after the construction of Hirakud dam in 1957, various industries were established in the adjacent area of dam in order to exploit the huge deposited mineral resources of the region. Even if the dam had no stated objective of giving water to industries, the government of Orissa had granted permission to six industries, collectively were drawing 30.38 cusecs of water. But the conflict became more intense, when the Government of Orissa again gave permission to draw 330.38 cusecs of water to various industries soon after the economy was liberalized during the 1990s. This led to several protests and movements by farmers of the locality against the government decision; however, the movement was suppressed by using mighty police force.

It has shown from the above study that Hirakud dam has not fulfilled any of its objectives. It has failed to manage floods, well short of target in terms of hydropower production and irrigation. Navigation remained a project of distant dream. It brought catastrophic environmental and ecological changes and uneven development in the region. Now, irrigation is being sacrificed for industrial profits. The people, who got displaced for the sake of national interests were not compensated and rehabilitated, treated like abandoned people. Various oppositions were made against the proposition of Hirakud dam. In the report of 1946, construction of dam at Hirakud was opposed by the engineer M.G. Rangaiya, opined that the objectives mentioned in the original plans will not be fulfilled, if the dam is constructed at Hirakud. Today, after 50 years of construction, the Hirakud dam is facing the exact situations, what he had predicted. But, it was unfortunate that the government didn't listen to his suggestions, and it implemented the TVA model of USA without any proper technical feasibility study and the socio-economic implications of involuntary displacement in the region.

Although the nature of the domestic opposition to the Hirakud dam was strikingly similar to domestic anti-dam struggle that emerged in the greater numbers in India during the 1970s, it failed to prosper due to the absence of NGOs, transnational allied advocacy networks, legitimized global norms on human rights, indigenous peoples and the environmental lobbies. The domestic resistance to Hirakud dam was politically too weak to influence the institution of India's democratic regime to either halt or even reform the Hirakud project. Besides the failure to attract larger alliances, the resistance to Hirakud dam was overshadowed by the nationalistic rhetoric of nation building in these early years of post-Independence. In the early years of Independence, the Prime Minister of India, Jawaharlal Nehru had, in fact once said 'Dams are temples of modern India'.

In this context, examining the linkage between TVA of USA and the MPRVD on the River Mahanadi shows that, the project of Hirakud dam was blindly executed without any proper study and its socio-economic impacts. Although various studies on flood management possibility of river Mahanadi rejected the idea of MPRVD in the early years, those rejections were put aside and ignored in the name of nationalism in the post-Independence era.

The study of World Commission on Dams Report (2000) found that large dams display a high variability in delivering predicted performances and related social benefits with a considerable falling short of physical and economic targets. It is also less profitable in economic terms and has a range of extensive socio-economic and environmental impacts. Thus, the true profitability of these schemes remains elusive. A close look at the performance of Hirakud dam clearly shows that it has underperformed in every respect and has had large-scale socio-economic impacts.

Given these circumstances, the question arises that should large dams be built? The unfortunate answer is yes. Large dams remain a necessary development option for providing water and energy resources to populations in developing countries that are in crisis. Taking the case of India, its growing population and rising need for food and water present a difficult situation. Further, it is compounded by dwindling monsoons that concentrate rainfall within a relatively short time period and in such situation, it is hard to see an alternative in some large storage reservoirs.

M.G. Rangaiya opined that Naraj was the right place to construct dam rather than at Hirakud. Flood management could be possible at Naraj by making dam of 65 feet height. Dam at Hirakud would not check the flood if heavy rain occurs. Irrigation and generation of electricity could not be possible if there was scanty and medium rainfalls. Therefore, it is recommended here that dams should only be constructed after the 'best practices' options assessment process that gives sufficient emphasis to environmental and social issues and only where adequate policies exists and implemented and where project authorities, contractors and consultants have a legal responsibility to follow whatever contractual conditionalities are necessary to implement the project as intended. Besides, dams should not be constructed in the most favourable zone of matrix that gives extensive weightage to only high technical advantages. The process of dam building should be based on a thorough analysis of alternatives. Efforts for alternative dam site should be identified on the same river to fulfil the same objectives with least social, cultural and environmental impacts.

Large areas of lands were acquired for the purpose of Hirakud dam, however, lands acquired for the purpose of offices, residential colony for the dam officials, reservoir power houses and the irrigation canals had not been assessed in the baseline survey. Thus, lots of people got indirectly affected. Compensations to those displaced people were very late by the Government of Orissa. They did not get their compensation in right time and many of the cases are still in controversy. The omission of such indirect impacts has been embroiled in controversy. Resettlement planning must focus not only on directly affected by the reservoir, but also identify the indirectly affected, caused by its associated infrastructure and auxiliary facilities.

Hirakud dam completely submerged 116 villages and partially affected 133 villages. Among the 133 partially affected villages, 48 villages houses submerged but the cultivable land and some houses remained outside. In the rest 85 villages lands were submerged but the houses remained outside. But, ironically they were not given proper compensations and deprived of rehabilitation. Therefore, emphasis needs to be given on the partially displaced people who often remain marginalized from compensations and rehabilitation programme.

The displaced people of Hirakud dam were resettled in 18 rehabilitation camp by the Government of Orissa. However, in reality, people were staying in 16 rehabilitation camps and other two rehabilitation camps are still lying vacant. Even in those 16 rehabilitation camps, there is sparse population, less than its intended capacity and many people left the rehabilitation camps after staying few years due to the crisis of livelihood and other infrastructural problems. Systematic stakeholders' involvement acts like self correcting mechanisms that enable timely identification of problems and helps in finding solutions for them. The key stakeholders in the resettlement programme are the affected people, their representatives, host populations, NGOs or other civil society organizations working in the area, local governments of the affected and the resettlement areas, project contractors, funding agencies, consultants conducting various studies and engineering and resettlement units in the project design and implementation agencies. Therefore, resettlement institutions need to have multi-sectoral involvement conducting a diverse range of activities, such as land taking, impact measurement, physical relocation, job provision, land development, credit provision and training to ensure livelihoods. The most important factor that needs to be addressed, both in selecting a particular dam for construction and facilitating smooth implementation of resettlement is the institutional capacity of the resettlement implementation agencies.

Moreover, the impoverishment risks of the oustees of Hirakud dam could have been mitigated by consultation and negotiations by the government of Orissa and the displaced communities and arguably the relocation would not be involuntary at all and no liberty would have been lost. It has been observed that the government didn't consult those whose vital interests were at stake. For oustees, giving them compensations in terms of monetary value for their productive assets (submerged cultivable lands) is unjust. The government could have resettled and rehabilitated in the irrigated area of Hirakud dam by giving them irrigated land, so that they would not have lost their sources of income. On the other hand, it would have also revived their bases of self-respect.

Contextualizing the Rawlsian perspective here, the idea of compensation is an idea of material replacement, like money for land, land for land, house for house, etc. is simply a bad idea and unjust. It is largely due to the fact that the experience of material replacement and rehabilitation still leaves people worse off. Rather, there should be the replacement of lost opportunities. His general conception of justice goes beyond compensation for economic losses. First, the displaced should not be worse-off in assets and resources by any development projects and on the contrary, it needs to be advantageous to them. Second, in the process of displacement and resettlement, there should be free negotiation between the state and community and all members of the communities need to be fairly represented in order to find out the best possible solutions to restore equality and justice. On the other hand, it also gives communities a legitimate right to refuse displacement, if it imposes disadvantages on other communities. Third, the social and cultural means of self-respect of the displaced communities should not be damaged in the process of relocation. The basis of self-respect need to revive and reinforce the social networks and associations among their members. Dams, referring to Rawlsian conception of justice, can only be claimed as development and justifiable when all the displaced people get their due without losing their income, wealth, basis of self-respect and social goods.

Soon after the construction of the dam, its catchment, reservoir and command have undergone considerable economic, environmental and ecological changes. The increasing number of industries around the reservoir and mining activities along the river basin is not only polluting the river water but also leading to number of conflicts between farmers and industries. Government must promulgate such kind of policies, which should be protective to environment and ecology. The decreasing trend of areas of irrigation in case of Hirakud dam has brought massive unrest among farmers. It is due to the increasing withdrawal of water for industrial use from the Hirakud reservoir. Besides it, the lack of maintenance of irrigation canals and government changing policies towards agriculture has been worsening the situation further. Therefore, government should give first priority to agriculture instead of industries. The water policy of the government should be pro farmer rather than pro industries. There should be heavy investment in secondary and tertiary canal system and drainage system which would accelerate the surface irrigation systems attached to large dams.

Plates

Hirakud Dam during the Period of Construction (1949-1957)

Hirakud Dam during the Period of Construction (1949-1957)

Completion of Hirakud Dam Project in August 1956

Hirakud Dam inaugurated by Pt. Jawaharlal Nehru on 13 January 1957

Hirakud Dam's First Hydropower Plant Commissioned on 18 December 1956

Irrigation of Hirakud Dam First Started on 7 September 1956

References

ACHR Weekly Review (2007), 'India Failed National Rehabilitation and Resettlement Policy, 2007', Accessed on 20 July 2008, URL: http://www.achrweb.org/review/2007/198-07.html.

Adams, William (2000), 'The Social Impact of Large Dams: Equity and Distributional Issues', Thematic Review, Social Issues I.1, Prepared for the World Commission on Dams.

Agarwal, G.D. (1967), *The Story of the Hirakud Dam Project*, Delhi: Ministry of Irrigation and Power, Government of India.

Altinbilek, Dogan (2002), 'The Role of Dams in Development', *Water Resource Development*, Vol. 18, No. 1, pp. 9–24.

Asian Development Bank (2003), 'Gender Checklists: Resettlement', Accessed on 6 February 2007, URL: http://www4.worldbank.org/afr/ssatp/resources/html/gender-rg/source%20%20documents/tool%20kits%20&%20guides/designing%20projects/tlpro9%20resettlement%20checklist%20adb%2003.pdf.

Asian Indigenous Tribal People Network (2006a), 'A Critique of India's Draft National Rehabilitation Policy', Indigenous Issues, Occasional Briefing Papers, New Delhi, Accessed on 24 August 2007, URL: http://www.aitpn.org/issues/II-08-06-rehabilitation.pdf.

—— (2006b), 'The Orissa Rehabilitation and Resettlement Policy, 2006: An Instrument of Further Displacement', Indigenous Issues, Occasional Briefing Papers, New Delhi, Accessed on 27 August 2006, URL: http://www.aitpn.org/issues/ii-07-06-orrp.pdf.

—— (2006c), 'Déjà vu: One Year after the Kalinganagar Massacre', *Indigenous Rights Quarterly*, New Delhi, Accessed on 27 August 2006, URL: http://www.aitpn.org/irq/vol-ii/story03.htm.

ASME Digital Collection (2012), 'Hydropower Generation in India: Status and Challenges', Accessed on 14 August 2012, URL: http://ebooks.asmedigitalcollection.asme.org/content.aspx?bookid=312&sectionid=38782244.

Baboo, Balgovind (1991), 'State Policies and People's Response: Lesson from Hirakud Dam', *Economic and Political Weekly*, Vol. XXVI, No. 4 12 October, pp. 2373–2379.

Bandhopadhyay, J., B. Mallik, M. Madal and S. Perveen (2002), 'Dams and Development: Report on Policy Dialogue', *Economic and Political Weekly*, Vol. 37, No. 40, 5–11 October, pp. 4108–4112.

Bartolome, Leopoldo Jose, Chris de Wet, Harsh Mander and Vijay Kumar Nagraj (2000), 'Displacement, Resettlement, Rehabilitation, Reparation and Development', Thematic Review, Social Issue 1.3, Prepared for the World Commission on Dams.

Baxi, Upendra (1989), 'Constitutional and Legal Aspects of Rehabilitation and Displacement', in Enakshi Thukral and Walter Fernandes (eds), *Development, Displacement and Rehabilitation*, New Delhi: Indian Social Institute.

Benjamin, N. (1971), 'Cauvery Waters Dispute', *Economic and Political Weekly*, Vol. 6, No. 34, pp. 1794–1795.

Bharati, M. and S. Rao (1999), 'Linking Development to Displacement', *Economic and Political Weekly*, Vol. 34, No. 22, 29 May, pp. 1374–1375.

Bhatia, Ramesh (2007), 'Water and Economic Growth', in John Briscoe and R.P.S. Malik (eds), *Handbook of Water Resources in India: Development, Management and Strategies*, New Delhi: Oxford University Press.

Biju, M.R. (2015), 'Land Acquisition Amendment Bill, 2014: U-Turn by the BJP', *Mainstream Weekly*, Vol. LIII, No. 3, August 8 (accessed from URL: http://www.mainstreamweekly.net/article5868.html).

Biswal, Umesh and Rabisankar (2007), 'Surendra Sai, Kisan Pattanayak o Prasanna Pandanka Matire Kampaniku Pani Dele Jalajuddha Heba', *Samadrusti* (Oriya), Vol. 1, No. 15, 16–31 January, pp. 17–18.

Bradshaw, Michael (1984), 'TVA at Fifty', *Geography*, Vol. 69, No. 3, June, pp. 209–220.

Centre for Science and Environment (1999), *The Citizens' Fifth Report: National Overview*, New Delhi.

Cernea, Michael M. (1988), 'Involuntary Resettlement in Development Projects', World Bank Technical Paper No. 80, Washington, D.C.: World Bank.

—— (1995), 'Understanding and Preventing Impoverishment from Displacement: Reflections on the State of Knowledge', *Social Action*, Vol. 45, April–June, pp. 261–275.

— (1998), 'Socio-Economic and Cultural Approaches to Involuntary Population Resettlement', Technical Paper, Washington, D.C.: World ank.

(2000), 'Risks, Safeguards and Reconstructions: A Model for ulation Displacement and Resettlement', *Economic and Political kly*, Vol. 35, No. 41, 7–13 October, pp. 3659–3677.

Michael M. and Christopher McDowell (eds) (2000), *Risk and truction: Experiences of Resettlers and Refugees*, Washington, orld Bank.

Srinivas (2012), 'Disputes, (De) Politicization and Democracy: River Disputes in India', Working Paper No. 108, RULNR per No. 13, January.

Choudhury, Pranab, Jinda Sandbhor and Priyabrata Satapathy (2012), *Floods, Fields and Factories: Towards Resolving Conflicts around the Hirakud Dam*, Pune: Forum for Policy Dialogue on Water Conflicts in India and Odisha State Resource Centre.

Clapp, Gordon R. (1948), 'TVA Learn as it Teaches', *The Phi Delta Kappan*, Vol. 29, No. 5, January, pp. 235–237.

D'Souza, Rohan (2006), *Drowned and Dam: Colonial Capitalism and Flood Control in Eastern India*, New Delhi: Oxford University Press.

Dankalmair, Christine and Leopoldo Bartolome (1999), 'The Experience with Dams and Resettlement in Argentina', Contributing Paper, Prepared for Thematic Review, Social Issue 1.3: Displacement, Resettlement, Rehabilitation, Reparation and Development, World Commission on Dams.

Datt, Ruddar and K.P.M. Sundaram (2003), *Indian Economy*, New Delhi: S. Chand & Company Ltd.

Deng, Francis M. (1998), *Internally Displaced Persons: Complications and Analysis of Legal Norms*, New York and Geneva: United Nations.

Deudney, Daniel (1981), 'Rivers of Energy: The Hydropower Potential', Worldwatch Paper 44, Washington, D.C.: Worldwatch Institute, June.

Drydyk, Jay (1999), 'Development-induced Displacement and Rawls' "General Conception" of Justice', Paper for Project on Economic Policy, Population Displacement and Development Ethics, Canada: York University.

Fernandes, Walter (1997), 'The Draft National Policy for the Rehabilitation of DPs: Is the Displacement Inevitable', in Walter Fernandes and Vijay Paranjpya (eds), *Rehabilitation Policy and Law in India: A Right to Livelihood*, New Delhi: Indian Social Institute.

—— (2004), 'Rehabilitation Policy for the Displaced', *Economic and Political Weekly*, 20 March, pp. 1191–1193.

Fernandes, Walter (ed.) (1999), *The Land Acquisition (Amendment) Bill 1998: For Liberalization or for the Poor?*, New Delhi: Indian Social Institute.

Fernandes, Walter and Enakshi Ganguly Thukral (eds) (1989), *Development, Displacement and Rehabilitation*, New Delhi: Indian Social Institute.

Fernandes, Walter and Vijay Paranjpya (1997), 'Hundred Years of Involuntary Displacement in India: Is the Rehabilitation Policy an Adequate Response?', in Walter Fernandes and Vijay Paranjpya (eds), *Rehabilitation Policy and Law in India: A Right to Livelihood*, New Delhi: Indian Social Institute.

Government of India (2008), *Development Challenges in Extremist Affected Areas: Report of an Expert Group*, New Delhi: Planning Commission.

—— (1947), 'Mahanadi Valley Development: The Hirakud Dam Project', Central Waterways Irrigation and Navigation Commission, Government of India Press, Simla.

Government of India (2013), 'The Right to Fair Compensation and Transparency in Land Acquisition, Resettlement and Rehabilitation Act, 2013' *The Gazette of India,* Ministry of Law and Justice, pp. 1–45.

Government of India (2014), 'The Right to Fair Compensation and Transparency in Land Acquisition, Resettlement and Rehabilitation (Amendment) Ordinance, 2014' *The Gazette of India,* Ministry of Law and Justice, pp. 1–4.

Government of Orissa (2006a), 'Flood Report of Hirakud Reservoir for 2006', Department of Water Resources, Superintendent Engineer Hirakud Dam Circle, Burla.

—— (2006b), 'Report of the Technical Expert Team on the Memorandum Submitted by Hirakhand Nagarik Parishad', Department of Water Resources.

—— (2006c), 'Orissa Resettlement and Rehabilitation Policy 2006', *The Orissa Gazette,* 14 May.

—— (2007a), 'Hirakud Cell Status Report' (HLO), Collectorate Office, Sambalpur, April.

—— (2007b), *State Water Policy,* Ministry of Water Resources, Bhubaneswar, March.

Green, C.H., D.J. Parker and S.M. Tunstall (2000), 'Assessment of Flood Control and Management Options', Prepared for Thematic Review, Option Assessments, IV.4, Prepared for the World Commission on Dams.

Gritzinger, Douglas (1987), 'New Lands Irrigation: Promise and Panacea?', *MERIP Middle East Report,* No. 145, March–April.

Gururaja, Srilakshmi (2000), 'Gender Dimensions of Displacements', *Forced Migration Review,* Vol. 9, December, pp. 13–16.

Hemadri, Ravi (1999), 'Dam, Displacement, Policy and Law in India', Contributing Paper, Prepared for Thematic Review, Social Issue 1.3, Displacement, Resettlement, Rehabilitation, Reparation and Development, World Commission on Dams.

Heming, Li, Paul Waley and Phil Rees (2001), 'Reservoir Settlement in China: Past Experience and the Three Gorges Dam', *The Geographical Journal,* Vol. 167, No. 3, September, pp. 195–212.

Hirakhand Nagarik Parishad (2007), *Smaranika 2007,* Annual Report, Sambalpur.

Huyck, Earl E. (1953), 'The Colombo Plan: Progress on the Sub-Continent', *Middle East Journal*, Vol. 7, No. 1, pp. 88–99.

India Hydro Energy, Accessed on 14 August 2012, URL: http://www.eai.in/ref/ae/hyd/hyd.html.

Inter-American Development Bank (1998), *Involuntary Resettlement: Operational Policy and Background Paper*, Washington, D.C., October, No. IND – 103, Accessed on 10 April 2007, URL: http://idbdocs.iadb.org/wsdocs/getdocument.aspx?docnum=362109.

International Commission on Large Dams: 'Dams and World Water', Accessed on 10 August 2012, URL: http://ussdams.com/ussdeducation/bookfory.pdf.

International Labour Organisation: 'Convention No. 169', Accessed on 3 December 2012, URL: http://www.ilo.org/indigenous/conventions/no169/lang–n/index.htm.

International River Network (2005), *Financing Dams in India: Risks and Challenges*, Accessed on 17 July 2007, URL: http://www.irn.org/pdf/india/financingdams2005_text.pdf.

Iyer, Ramaswamy R. (2001), 'Water: Charting a Course for the Future-1', *Economic and Political Weekly*, 31 March, pp. 1115–1122.

Jena, Manipadma (1998), 'Development and Deprivation', *Economic and Political Weekly*, Vol. 43, No. 20, 11 April, pp. 821–822.

—— (2006), 'Draft Resettlement and Rehabilitation Policy, 2006', *Economic and Political Weekly*, 4 February, pp. 384–387.

Jing, Jun (1999), *Displacement, Resettlement, Rehabilitation and Development – China Report*, Contributing Paper, Prepared for Thematic Review, Social Issue 1.3: Displacement, Resettlement, Rehabilitation, Reparation and Development, World Commission on Dams.

Joyce, Stephanie (1997), 'Is It Worth a Dam?', *Environmental Health Perspectives*, Vol. 105, No. 10, October, pp. 1050–1055.

Karan, P.P. (1994), 'Environmental Movements in India', *Geographical Review*, Vol. 84, No. 1, January, pp. 32–41.

Khagram, Sanjeev (2004), *Dams and Development: Transnational Struggle for Water and Power*, New Delhi: Oxford University Press.

Klingensmith, Daniel (2007), *'One Valley and a Thousand': Dams, Nationalism and Development*, New Delhi: Oxford University Press.

Kujur, Joseph Marianus (2005), 'A Tribal Reading of the Resettlement and Rehabilitation Policy, 2003', *Social Action*, Vol. 55, April–June, pp. 138–151.

Loescher, Gil and James Milner (2005), 'Protracted Refugee Situations: Domestic and International Security Implications', Adelphi Paper No. 375, London: Routledge, July.

Mahalanobis, P.C. (1940), 'Rain Storms and River Floods in Orissa', *Sankhya: The Indian Journal of Statistics*, Vol. 5, No. 1, pp. 1–20.

Mahalik, Nirakar (2005), 'A Dam Across Mahanadi: A Dream Project of Dr. A.N. Khosala', *Orissa Review*, April, Accessed on 3 June 2007, URL:

http://orissagov.nic.in/e-magazine/orissareview/apr2005/englishpdf/damacrossmahanadi.pdf.

Mahapatra, L.K. (1991), 'Development for Whom? Depriving the Dispossessed Tribals', *Social Action*, July–September, pp. 271–287.

Manav Adhikar Seva Samiti (2006), 'Hajuchhi Hirakud–Tejuchhi Dwanda', *Panira Dagara* (Oriya Journal), May–June, pp. 4–11.

McCartney, M.P., C. Sullivan and M.C. Acreman, 'Dams, Ecosystem Functions and Environmental Restoration', Prepared for Thematic Review II.1, Accessed on 21 August 2012, URL: http://www2.wii.gov.in/eianew/eia/dams%20and%20development/kbase/contrib/env244.pdf.

McCully, Patrick (2001), *Silenced Rivers: The Ecology and Politics of Large Dams*, London and New York: Zed Books.

Meher, Rajkishor (2011), 'Big Dam, Big Failures: A Study of the Canal Irrigation System and the Deprived Tail-End Farmers in the Hirakud Command Area of Orissa, India', *Journal of Asian and African Studies*, June 24, pp. 422–432.

Mehta, Pooja (2005), 'Internally-Displaced Persons and the Sardar Sarovar Project: A Case for Rehabilitative Reform in Rural Media', *American University International Law Review*, Vol. 20, No. 3, pp. 615–647.

Menon, M.S. (2003), 'Sardar Sarovar Project: Another Perspective', *Economic and Political Weekly*, 27 September, pp. 4095–4096.

Milewski, Joseph, Dominique Erge and V. Roquet (1999), 'Dams and Benefit Sharing', Contributing Paper, Prepared for Thematic Review I.1: Social Impacts of Large Dams Equity and Distributional Issues, World Commission on Dams.

Ministry of Rural Development (2004), 'National Policy on Resettlement and Rehabilitation for Project Affected Families', *Gazette of India*, Extraordinary Part 1, Section 1, No. 46, 17 February 2004.

Mishra, A. and M.C. Dash (1984), 'Desertification Around Hirakud Reservoir', *The Environmentalist*, Vol. 4, No. 1, pp. 51–58.

Mohile, D. Anil (2007), 'Development Policies and Programmes', in John Briscoe and R.P.S. Malik (eds), *Handbook of Water Resources in India: Development, Management and Strategies*, New Delhi: Oxford University Press.

Mohsin, Amena (1997), *The Politics of Nationalism: The Case of Chittagong Hill Tracts, Bangladesh*, Dhaka: University Press Limited.

Morgan, Leslie (1952), 'River Development in India', *Far Eastern Survey*, Vol. 21, No. 8, 21 May, pp. 82–83.

Murty, G.N. and P.K. Mishra (1994), 'Performance Overview of Hirakud Dam Project: A Case Study', Fifth National Water Convention, Organized by National Water Development Agency, Government of Haryana and Indian Water Resources Society, 25–27 February.

Narayanamoorthy, A. and L. Venkatachalam (2011), 'Farmers' Right to Water', *Business Line*, September, Accessed on 15 December 2012, URL:

http://www.thehindubusinessline.com/opinion/article2430020.ece?homepage=true.

Nayak, Ranjit (2000), 'Risks Associated with Landlessness: An Exploration towards Socially Friendly Displacement and Resettlement', in Michael M. Cernea and Christopher McDowell (eds), *Risk and Reconstruction: Experiences of Resettlers and Refugees*, Washington, D.C: World Bank.

Orissa Hydropower Corporation Ltd. (2007), *Hirakud Power System: Project Features*, Accessed on 12 July 2007, URL: http://www.ohpcltd.com/hirakud/project-feature.htm.

Padel, Felix and Samarendra Das (2010), *Out of this Earth: East India Adivasis and the Aluminum Cartel*, New Delhi: Orient Blackswan.

Pandey, Balaji (1998), *Depriving the Underprivileged for Development*, Bhubaneswar: Institute for Socio-Economic Development.

Patel, Vibhuti (2010), 'Human Rights Movements in India', *Social Change*, Vol. 44, No. 4, pp. 459–477.

Pathak, Sekhar (2005), 'Tehri Dam: Submersion of a Town, Not of an Idea', *Economic and Political Weekly*, Vol. 40, No. 33, pp. 3637–3639.

Pati, Bikash and Manas Biswal (2009), 'Hirakud Dam: Fifty Mournful Years', *Dams, Rivers and People*, June–August, pp. 7–11.

Pattanaik, S.K., B. Das and A. Mishra (undated), 'Hirakud Dam Projects: Expectations and Realities', in Edward Goldsmith and Nicholas Hildyard (eds), *People and Dams*, Wadebridge, UK: Wadebridge Ecological Centre.

Patwardhan, Amrita (1999), 'Dams and Tribal People in India', Contributing Paper, Prepared for Thematic Review 1.2: Dams, Indigenous People and Vulnerable Ethnic Minorities, World Commission on Dams.

Ramanathan, Usha (1995), 'Displacement and Rehabilitation: Towards A National Policy', *Lokayan Bulletin*, Vol. 11, No. 5, pp. 41–56.

Ramdas, N. (2001), 'How to Dams Nature's Fury', *The Hindu*, Saturday, 21 July, Accessed on 8 July 2007, URL: http://www.hindu.com/thehindu/2001/07/21/stories/0221000e.htm.

Rao, V.K.R.V. (1952), 'India's First Five Year Plan: A Descriptive Analysis', *Pacific Affairs*, Vol. 25, No. 1, March, pp. 3–23.

Rawls, John (1971), *A Theory of Justice*, Cambridge: Harvard University Press.

Robinson, Scott (1999), 'The Experience with Dams and Resettlement in Mexico', Contributing Paper, Prepared for Thematic Review, Social Issue 1.3: Displacement, Resettlement, Rehabilitation, Reparation and Development, World Commission on Dams.

Samadrusti Televisions (2007), *Pani* (Oriya Documentary Movie), Bhubaneswar.

Samaj (2007), 'Hirakud Jalabhandarara Khyamatabrudhi Eka Abantar Parikalpana', Sambalpur Edition, 11 November, p. 3.

Sambada (2007), 'Pulishra Lathi Charge, Sambadikanaka Sameta Chalishru Addhika Aahata', Sambalpur Edition, 7 November, p. 7.

—— (2007a), 'Hirakud Lathimada Ghatanare R.D.C. Tadanta', Sambalpur Edition, 8 November, p. 1 and 8.

—— (2007b), 'Silpaku Panidebu: Mukhyamantri', Sambalpur Edition, 27 November 27, p. 1.

Sanmuganathan, K. (2000), 'Assessment of Irrigation Options', Thematic Review IV.2, Prepared for the World Commission on Dams.

Saxena, K.B. (2015), 'The Ordinance Amending the Land Acquisition Law (2013): Farmers Lose out in the Unequal Contest of Power', *Social Change,* Vol. 42, No. 2, pp. 324–336.

Sekhar, A. (2007), 'Development and Management Policies: Perspectives of the Planning Commission', in John Briscoe and R.P.S. Malik (eds), *Handbook of Water Resources in India: Development, Management and Strategies*, New Delhi: Oxford University Press.

Sen, Sumit (1998), 'Exiled at Home: The International Regime of Internal Displacement', *Indian Journal of International Law,* Vol. 38, pp. 182–207.

Sharma, R.N. (2003), 'Involuntary Displacement: A Few Encounters', *Economic and Political Weekly,* Vol. 38, No. 9, 1–7 March, pp. 907–912.

Shylendra, H.S. (2002), 'Environmental Rehabilitation and Livelihood Impact: Emerging Trends from Ethopia and Gujarat', *Economic and Political Weekly*, Vol. 37, No. 31, 3–9 August, pp. 3286–3292.

Singh, Satyajit K. (1990), 'Evaluating Large Dams in India', *Economic and Political Weekly*, 17 March, pp. 561–574.

Sopher, David E. (1963), 'Population Dislocation in Chittagong Hills', *Geographical Review*, Vol. 53, No. 3, July, pp. 337–362.

South Asian Network on Dams, Rivers and People (1999), 'Assessment of Large Dams Options for Urban India: Large Dams have No Case', *Thematic Reviews: Options and Assessment*, Submitted to World Commission on Dams.

Sovani, N.V. and Nilakanth Rath (1960), *Economics of Multi-Purpose River Dam: Report of an Inquiry into the Economic Benefits of the Hirakud Dam*, Pune: Gokhale Institute of Politics and Economics and Bombay: Asia Publishing House.

Supakar, Karunakar (2004), *Itihaasra Parihaas* (Oriya), First Part, Sambalpur.

Tripathy, Surendra (2006), 'Kampaninku Pani Jogana Birodhare Pachimanchala Chhasinka Hirakud Band Dakhal Abhiyan', *Samadrusti* (Oriya Journal), Vol. 1, No. 11, 16–30 November, pp. 20–21.

UNHCR (1997–98), *The State of the World Refugees 1997–98: A Humanitarian Agenda*, Oxford: Oxford University Press.

United Nations Sustainable Development, 'United Nations Conference on Environment and Development', Rio de Janerio, Brazil, 3 to 14 June 1992, Agenda 21, Accessed on 2 December 2012, URL: http://www.un.org/esa/sustdev/documents/agenda21/english/agenda21.pdf.

Upadhayay, Sanjay and Videh Upadhayay (2002), *Handbook on Environmental Law, Volume I: Forest Laws, Wild Laws, Wildlife Laws and the Environment*, First Edition, New Delhi: Lexis Nexis.

Vakil, C.N. (1950), *Economic Consequences of Divided India: A Study of the Economy of India and Pakistan*, Bombay: Vora & Co. Publishers Ltd.

Wet, Chris de (1999), 'The Experience with Dams and Resettlement in Africa', Contributing Paper, Prepared for Thematic Review, Social Issue 1.3: Displacement, Resettlement, Rehabilitation, Reparation and Development, World Commission on Dams.

White, Gilbert F. (1957), 'A Perspective of River Basin Development', *Law and Contemporary Problems*, Vol. 22, No. 2, pp. 157–184.

Wikipedia, 'Electricity Sector in India', Accessed on 14 August 2012, URL: http://en.wikipedia.org/wiki/electricity_sector_in_india.

—— 'Damodar Valley Corporation', Accessed on 28 August 2012, URL: http://en.wikipedia.org/wiki/damodar_valley_corporation.

—— 'Upper Wardha Dam', Accessed on 28 August 2012, URL: http://en.wikipedia.org/wiki/upper_wardha_dam.

—— 'Western Cape Water Supply System', Accessed on 19 August 2012, URL: http://en.wikipedia.org/wiki/western_cape_water_supply_system.

Wood, John R. (1993), 'India's Narmada River Dams', *Asian Survey*, Vol. 33, No. 10, October, pp. 968–984.

—— (2007), The Politics of Water Resource Development in India: The Narmada Dams Controversy, New Delhi: Sage Publications.

World Bank (1998), 'Involuntary Resettlement: Operational Policy and Background Paper', No. IND – 103, October, Washington, D.C.: World Commission on Dams.

—— (2000), 'Involuntary Resettlement: The Large Dam Experience', *Precis*, No. 194, Operations Evaluation Department, Winter.

World Commission on Dams (2000), *Dams and Development: A New Framework for Decision Making*, London: Earthscan Publications.

World Health Organization (1999), 'Human Health and Dams', Submission by the World Health Organization to the World Commission on Dams, Geneva, November.

Zhang, Lubiao (1999), 'Social Impact of Large Dams: The China Case', Contributing Paper, Prepared for Thematic Review I.1: Social Impacts of Large Dams Equity and Distributional Issues, World Commission on Dams.

Index